William J. Spillman
and the Birth of
Agricultural Economics

MISSOURI BIOGRAPHY SERIES
WILLIAM E. FOLEY, EDITOR

William J. Spillman

and the Birth
of Agricultural
Economics

Laurie Winn Carlson

University of Missouri Press
Columbia and London

Copyright © 2005 by
The Curators of the University of Missouri
University of Missouri Press, Columbia, Missouri 65201
Printed and bound in the United States of America
All rights reserved
5 4 3 2 1   09 08 07 06 05

Library of Congress Cataloging-in-Publication Data

Carlson, Laurie M., 1952–
William J. Spillman and the birth of agricultural economics /
Laurie Winn Carlson.
   p.   cm. — (Missouri biography series)
Summary: "Biography of William J. Spillman, scientist and educator for the United
States Department of Agriculture. Explores Spillman's role in the development of the
agricultural economics, the agricultural New Deal, genetics research, agricultural
education and the Cooperative Extension Service, the post–World War I overproduction crisis, and the Law of Diminishing Returns"—Provided by publisher.
Includes bibliographical references and index.
ISBN 0-8262-1581-5 (alk. paper)
   1. Spillman, W. J. (William Jasper)   2. Agricultural economists—Biography.
3. Agriculture—Economic aspects—United States—History.   4. Agricultural
education—United States—History.   5. United States Dept. of Agriculture—
History.   I. Title.   II. Series.
HD1771.5.S65C37   2005
338.1'092—dc22
2005000618

♾™ This paper meets the requirements of the
American National Standard for Permanence of Paper
for Printed Library Materials, Z39.48, 1984.

Designer: Stephanie Foley
Typesetter: Phoenix Type, Inc.
Printer and Binder: Thomson-Shore, Inc.
Typeface: Adobe Caslon

Contents

William Spillman standing in the test plots, or "grass garden," at Washington State Agricultural College and School of Science (now Washington State University) in Pullman, Washington. Beginning in 1894, he and students collected a wide variety of forage samples for experimentation. Here he is shown with a stand of reed canary grass *(Phalaris arundinacae)*. (William J. Spillman Papers, Manuscripts, Archives, and Special Collections, Washington State University)

Acknowledgments

Thank you to many who assisted me on this project, including Larry Stark, Trevor Bond, and Patsy Tate in Manuscripts, Archives, and Special Collections at the Washington State University Library and the staff at the University of Washington Library Archives. Joseph D. Schwarz, at the National Archives–College Park, and Sara Lee and Lynn Jones Stewart, at Special Collections of the National Agricultural Library, were very helpful. Thanks to Richard King at Vincennes University; Jane Ruth at Mann Library, Cornell University; William T. Stolz at the University Archives, University of Missouri–Columbia; Charles Mutschler and Elaine Zeiber-Breeding at the Eastern Washington University Archives; and Nancy Gale Compau of the Northwest History Room at the Spokane Public Library. Tim Steury, editor of *Washington State Magazine,* and Stephen Jones, the geneticist who fills Spillman's shoes at Washington State University today, provided insights and research direction. Thanks as well to Larry Clark of the Washington State Grange; Robin Henig, author of *Monk in the Garden;* John Perkins of Evergreen State College; Ann Greene at the University of Pennsylvania (a friend and colleague in agricultural history); and Mike Green, Eastern Washington University history professor emeritus, who kept encouraging from the sidelines. I appreciate the assistance from the faculty and staff of the History Department at Washington State University, especially Pat Thorsten and Dwayne Dehlbom, who keep everything functioning, and Roger Schlesinger, Paul Hirt, Michael Egan, David Coon, Leroy Ashby, and Susan Armitage for valuable advice and expertise.

Thanks as well to Annette Wenda, Jane Lago, Clair Willcox, and the staff at the University of Missouri Press.

And, thanks to Terry.

William Spillman in 1894, while living in Monmouth, Oregon, where he taught at the Oregon State Normal School. He left that year for a position at the Washington State Agricultural College and School of Science in Pullman, Washington. (William J. Spillman Papers, Manuscripts, Archives, and Special Collections, Washington State University)

William J. Spillman
and the Birth of
Agricultural Economics

WHEAT HYBRIDS
Experiment Station

Spillman's wheat hybrid research yielded improved varieties and also helped him recognize and explain the concepts involved in Mendelian genetics. This display is likely the same one he presented at the American Agricultural Colleges and Experimental Stations meeting in Washington, D.C., in 1901. His explanation both proved Mendel's genetics theories correct and established Spillman as an American codiscoverer of Mendel. After the presentation, he was hired by the USDA and embarked on a career in national agricultural research and policy. (William J. Spillman Papers, Manuscripts, Archives, and Special Collections, Washington State University)

Introduction

William Jasper Spillman is an elusive figure. Although his name is recognized by scientists and economists, the man and his life have faded from memory. Washington State University, in Pullman, Washington, recognizes his contributions made there in the 1890s with an experimental farm, the Spillman Farm, as well as with an annual wheat exhibition tour, the Spillman Farm Tour. A bronze plaque commemorating Spillman hangs in the foyer of Johnson Hall, one of the agriculture classroom buildings on campus, but few students or faculty know anything about the man. Therefore, when my dissertation adviser, David Coon, suggested I consider the man as a potential research subject, I hesitated because I had never heard of him. A trip to the Manuscripts, Archives, and Special Collections division of Holland Library at Washington State University gave me an overview: Spillman had come to the school just as it was being launched as a state landgrant agricultural college and had indeed developed a variety of wheat that improved farmers' yields as well as the credibility of the new agricultural college. But he was at Pullman for only a few years, so I did not see a lot of potential in looking at his life. Plant genetics in eastern Washington in 1900 did not seem to offer much in the way of historical questions and answers; I sought a larger topic—something with conflict, triumph, loss, even corruption and vice, elements that make history intriguing and worth the writing.

So when I was thumbing through a copy of Upton Sinclair's critique of higher education, *The Goose-Step: A Study of American Education,* I stopped short at Sinclair's comment: "One of the government employees who is not a corporation hireling is Professor W. J. Spillman, chief of the Bureau of Agricultural Economics, and editor of a farm paper."[1]

1. Sinclair, *Goose-Step,* 198.

Sinclair went on to describe Spillman's resistance to the Rockefeller Foundation's efforts to control higher education. Sinclair's comments intrigued me, and I returned to the Spillman Papers in the archives and dug a little deeper to find out why that era's premier investigative journalist had lauded such an obscure individual.

The archives held a typescript written by Ramsay Spillman, William's son, in the late 1930s, after his father's death. It was a biography of his father. Ramsay Spillman was forty-nine years old at the time, a respected physician and medical writer. In 1937, he translated German physician J. Wagner-Jauregg's biography of Constantin von Economo, important because an epidemic of *encephalitis lethargica,* also known as Von Economo's disease, had stymied American medical experts. In 1927, Wagner-Jauregg had received the Nobel Prize in Medicine for his work on fever therapy for paresis. Ramsay had chosen a good subject for a scholarly project. His father's life, however, held great appeal, and Ramsay believed his manuscript would win the annual Atlantic Press Non-Fiction Contest. The manuscript did not win publication and the five thousand–dollar prize, and it is unknown whether Ramsay tried elsewhere to get it published. Another writer who entered the contest twice, winning five years earlier, in 1935, was Mari Sandoz, whose biography of her father became the classic *Old Jules.* Using Ramsay Spillman's manuscript, I began the search for William Jasper Spillman. He emerged as a man who bridged two eras in U.S. agriculture, a man who was both a modernizer and a critic of U.S. agriculture.[2]

Ramsay Spillman's family story begins with his grandfather Nathan Cosby Spilman (b. 1823) and his brother returning to the Midwest from the goldfields of California in 1850 with one thousand dollars each to start farms. Nathan purchased a two hundred–acre farm about seven miles north of Monett, Missouri (in the southwestern corner of Missouri, south of and midway between Springfield and Joplin). Nathan and his wife, Emily Paralee Pruit (b. 1830), had fifteen children; their son William, Willy, as he was called, was eleventh, born in 1863. In 1871, when Willy was seven years old, Nathan died from sunstroke while working in the fields. Ramsay Spillman recounts that he was remem-

2. Sandoz, *Old Jules,* 429. Von Economo's disease was the pandemic of 1916–1930, which I studied in my master's thesis, "Fits and Fevers: Salem's Witches and the Forgotten Epidemic" (Eastern Washington University, 1999); its cause remains a mystery to this day.

bered for being a loyal subscriber to the *Toledo Blade* newspaper, riding horseback ten miles every Saturday to pick it up, then reading it aloud to neighbors gathered at his farm to learn the news. He also subscribed to the popular pseudoscientific publication *Phrenological Journal,* which may have shaped young Willy's thinking about human behavior, science, and, in the future, eugenics. The *Phrenological Journal,* and similar publications, focused on a popularization of science as well as a holistic view of human health and success. Willy would continue to hold those views as he pioneered genetics and scientific breeding in his later career.[3]

We know little about Willy's early education other than that it was a traditional, small rural-school experience; he often later recited passages and rhymes from both McGuffey's readers and Noah Webster's blue-backed speller. At some point he began teaching in a small school near home.[4]

In 1881, seventeen-year-old Willy Spilman—clad in homespun trousers—left his teaching job and the family farm in Lawrence County, Missouri, and set out for college, hoping to attain his dream of someday owning a sheep ranch in eastern Oregon, where two of his older sisters had recently settled. He worked his passage to St. Louis as a cowpuncher on a train car of cattle sent to the stockyards. He took his clothing and twenty-three dollars saved up from teaching and farm labor.

From St. Louis he took the train to Columbia, where he enrolled at the University of Missouri. He found a job as a church janitor and took on corn-cutting jobs for a dollar a day on Mondays, a day when no classes met. To help pay his expenses at school, his brother John sent Willy his share of the money from the wheat crop on the family farm. "Since I have talked with students from other schools," he wrote his brother and sister, "I have concluded that the university is about as good a place for a poor boy, as can be found." He urged them to send their brother Nathan down and other "Lawrence County boys." He asked his siblings to buy out his share of the family farm so he could finish his education, become a surveyor, and head west. "You know there are five of us boys and nine shares so we all can't have two apiece, and one is not worth a great deal by itself."[5]

3. R. Spillman, "A Biography of William Jasper Spillman," 12.
4. Ibid., 16.
5. Ibid., 41.

Willy threw himself into college life, particularly enjoying mathematics and languages; he learned both French and German. He maintained his German knowledge by attending German-language church services and reading widely throughout the rest of his life. He maintained high grades and did well, particularly in elocution, gaining public speaking skills that would help his career later. While at college, he joined the local Grange chapter; "the Grange in this section is still alive," he wrote to his mother. He had been a member of Rural Grange no. 443 in Lawrence County and transferred his membership. He remained a Grange member until his death. After finishing his junior year, he took a job teaching algebra at Pierce City Baptist College, close to the family home, saving enough to return and finish his senior year.[6]

While at the university, Willy shifted to using initials, W. J., for William Jasper. He also modified his last name, changing the spelling from Spilman to Spillman, adding an extra *l* because his professors persisted in spelling it that way. His son, Ramsay, remarked that his father "was never one to let a fight for the right go by default, but he had no time to waste on what seemed to him to be trivialities." After all, in the scope of things, it was simply a matter of a consonant, and Willy—W. J.—had a larger vision for himself.[7]

His experience at the university did more than change how William Spillman spelled his last name; it provided the scientific and educational background necessary to move him from a barefoot farm boy to one of the nation's leading scientists, economists, and agricultural policy makers. He recognized the value of education early on, urging his brother and neighbors to come to college as well, a value he continued to hold the rest of his life. For William Spillman, pursuing knowledge and sharing that knowledge with others would become his lifework. His son, Ramsay, noted that the University of Missouri's impact on his father's life could not be overestimated: "The effect of the University on one who went there a farm laborer and emerged a potential scientist" was substantial.[8]

He graduated with a B.S. in 1886 and moved to Marshall, Missouri, where he taught high school for two years. He followed that with a

6. Ibid., 39.
7. Ibid., 40
8. Ibid., v.

position at Missouri State Normal School in Cape Girardeau, where he taught for three years. He met and married Mattie Ramsay (1865–1935), who was one of his students, in 1889. The couple moved to Indiana where Spillman taught botany and physics at Vincennes University and became active in the scientific community. It was here that he met Enoch A. Bryan, the university president who would later recruit him to Washington State. He subscribed to *Popular Science Monthly* for years and in 1889 was elected to membership in the Indiana Academy of Sciences after presenting two papers, "A Comparison of the Life Histories of Different Forms of Plants" and "The Height of the Atmosphere," at the group's annual meeting. Neither paper is extant. At the group's 1890 meeting, Spillman presented four papers, possibly his first formal scientific presentations. That same year, the University of Missouri awarded Spillman a master's degree in absentia, a practice common at the time to recognize meritorious work, effectively an honorary degree.[9]

In 1891, Spillman heard from one of his two older sisters who had relocated to Oregon that a vacancy existed for someone to teach science at the new Oregon State Normal School at Monmouth. Securing the job, the Spillmans loaded up their books and left the boardinghouse in Indiana, heading west. It was in Monmouth that the couple's only child, Ramsay, was born in 1891. An avid bicyclist, Spillman built a small wooden seat over the handlebars so he could take the child along on jaunts around campus and town. He taught chemistry, botany, and physiology and frequently took his students on collecting field trips. He joined a group of amateur naturalists and hikers, trekking twice to the top of Mount Hood, in 1892 and 1893. He wrote an article about the 1892 trip for the *Portland Oregonian,* "Above the Clouds: The Perilous Ascent of Mt. Hood. Detailed Description of the Dangers and Hardships and of the Pleasure That Awaits One at the Summit." That article appears to be his first published work.[10]

The panic of 1893 hit Oregon hard; the legislature failed to fund the normal school at Monmouth, and teachers survived on IOUs to town merchants. W. J. went to Salem to lobby for the school, where he discovered that a single legislator who was determined to destroy the school

9. Ibid., 112.
10. Ibid., 435.

held the crucial vote. "Just as the vote was being taken on the measure containing the appropriation," his son wrote, "Spillman sent in to the member in question a folded paper marked *Very Important. Read at once!*" Opening the paper, the legislator found an extremely illegible message and was thrown entirely off guard. "The legislator concentrated on the illegible contents to such a degree that the vote on the appropriation passed without his notice. The bill was carried." W. J. telegraphed to his wife, Mattie, "We got there. W. J. S." Monmouth had been saved.[11]

In 1894, his friend and colleague from Vincennes University, Enoch Bryan, was appointed president of the newly opened Washington State Agricultural College and School of Science in the small eastern Washington town of Pullman. Bryan offered Spillman a position teaching agriculture and horticulture. The salary was handsome for the time, eighteen hundred dollars per year, and in light of the financial problems in Oregon, he did not hesitate to accept Bryan's offer. Having studied neither subject in school, he agreed to accept the position only if someone else was hired to teach horticulture. At the same time, Spillman believed that his science background, as well as his childhood upbringing on the Missouri farm, made the agriculturist position a good fit. To better prepare for teaching agriculture, he spent several weeks at the University of Wisconsin observing agricultural teaching and research methods. He took a crash course in cheese making, familiarized himself with the newly created Babcock milk test that measured fat content in milk, and observed the Wisconsin livestock feeding program that was innovative for the time.

Arriving at the Washington State Agricultural College and School of Science, W. J. found a far different setting than the small diversified farms in his native Missouri or in western Oregon. The college was at the center of one of the world's premier wheat-growing areas, but in 1894, when W. J. arrived, the region was in distress: wheat had fallen to eighteen cents per bushel that year, on the heels of a very wet 1893 that left wheat rotting on the stem and sacks of swollen grain still abandoned on the ground. The national economy was in dire straits, too, and hundreds of Pacific Northwest men joined Coxey's Army of unemployed laborers headed eastward by train to demand government assistance.[12]

---

11. Ibid., 119.
12. See Carlos A. Schwantes, *Coxey's Army: An American Odyssey*.

W. J. set to work, establishing himself as a pragmatic and practical scientist who attempted to bring valuable information to the farmer. His first publication was *Silos and Ensilage,* because silos were new technology to the region. His first research project resulted in the bulletin *Feeding Wheat to Hogs,* to assist farmers with the rotting 1893 harvest and the low prices for the 1894 crop. Those publications, the result of careful scientific study, were only the first in what would be many attempts to use scientific means to solve practical problems.

Spillman's Role in Modern Agriculture

Spillman moved from his beginnings on a Missouri farm to become a leading agricultural scientist, educator, and administrator of agricultural policy in the modern age. As he applied science to agriculture during the first three decades of the twentieth century, he straddled several shifts in U.S. society: from populism to progressivism, from a largely rural nation to an urban one, from farming technology limited by folk knowledge to scientific agriculture, and from family farms to an industrial model based on international exports.

He "rediscovered" Mendel's laws of inherited characteristics through his research with wheat varieties better suited to eastern Washington farmers, laying a foundation for a U.S. field of genetic science. After joining the U.S. Department of Agriculture (USDA) under agriculture secretary James Wilson, he pioneered the field of farm management and helped establish federal educational programs for farmers that evolved into the Agricultural Extension Service. He was instrumental in organizing the field of agricultural economics, challenged existing precepts about farming with his explanation of the law of diminishing returns, and urged the government to intervene when postwar overproduction threatened to ruin rural economies. His book *Balancing the Farm Output,* published in 1927, laid the foundation for the New Deal and twentieth-century federal agricultural policy. Eventually, national agricultural policy reflected the concepts he developed regarding the need for federal involvement in agriculture and rural life.

A talented mathematician and scientist and a well-liked teacher and colleague, William Spillman did not always find the going easy. As it would be throughout the twentieth century, the Department of

Agriculture was awash in politics, and clashing ideas and interests intensified the demands on the department by industry and the public. Spillman resisted the heavy involvement by the Rockefeller Foundation within the department and urged a congressional investigation in 1914, which indirectly led to the creation of the Agricultural Extension Service based at agricultural colleges. The congressional probe was a turning point in his career from which he both gained notoriety as well as made enemies. Ramsay's memory of his father dwells largely on the issue, showing how important it was to his father's life. The memorials at his funeral service stress how difficult the challenge to corporate interests had been for him and how it had both affected Spillman's life as well as reinforced friends' and colleagues' respect and admiration for the man.

In many ways, William Jasper Spillman epitomized the Horatio Alger–type American success story. A Missouri farm youth who went off to state college because the family farm was too small for the many children, he used his education to make his way in the world, initially at an obscure western agricultural college and eventually at the center of national agricultural policy during the critical years of the Progressive Era. A devoted researcher and writer (with more than three hundred publications to his name), he did not become wealthy or very famous; rather, his story is that of the journey of a notable agricultural scientist, administrator, and educator through a pivotal era in the history of U.S. agriculture.

1

Scientific Breeding

W illiam Spillman's background in agriculture, his commitment to scientific methods, and his desire to work with farmers led him to identify one of the most elusive yet important scientific discoveries of the period. He not only creatively bred an improved strain of wheat but also identified how inherited characteristics passed on to the next generation. Spillman experimented with wheat varieties to suit the Palouse wheat-growing region, coming up with hybrid wheats while rediscovering Gregor Mendel's long-ignored laws of genetics. Spillman's work and its effects would have long-term effects on both wheat production and U.S. agriculture.

In the 1890s, as wheat breeders sought to create better crop varieties to suit North American climates, the field of genetics was in its infancy. Genetics, the biology of inherited characteristics, is a common topic now; the human genome project, cloning, and genetic testing are in the news daily. Although scientific breeding is largely a twentieth-century phenomenon, its origins reach back earlier. To eighteenth-century thinkers, inherited characteristics were a complete mystery. A leading European naturalist at that time, Georges-Louis Leclerc de Buffon, claimed that the male parent determined an offspring's extremities, whereas the female parent determined its internal parts and overall size and shape. Livestock breeders, like English farmer Robert Bakewell, began to change the shape of England's sheep, cattle, and horses through intensive inbreeding of closely related animals, but no one knew why animals (or plants) inherited particular characteristics. During the nineteenth century, things did not get much clearer. Ideas about inheritance included telegony, the belief that the first male to impregnate a female passed on inherited characteristics to all her offspring; the general acceptance of influences on fetal development due to sensory experiences of the mother during gestation; and the influence of astrology.[1]

Charles Darwin's groundbreaking work, *On the Origin of Species*, appeared in 1859, opening the frontiers of biological thought. Darwin tried to understand what caused individual differences, rejecting the accepted idea of the time that close inbreeding of progeny was how new breeds were formed. He argued that it was more of a smooth blending of parental characteristics in the offspring. He believed that variation between breeds was due to environmental influences, such as climate and food, on the reproductive system. Darwin believed too that genetic particles resided in the circulating fluids, passing through the entire body. His cousin Francis Galton put the idea to a test, transfusing blood between rabbits with different-colored fur. After breeding, however, the transfused blood failed to influence the fur color of the offspring. Yet Darwin's idea that variations between individuals were integral to the process of evolution by natural selection was a foundation for breeding work, even though it offered little useful practical application for breeders. His ideas offered "faith that variation could be mastered and controlled," and his work remained at the center of the field until the 1920s.[2]

Gregor Mendel, a chubby Austrian monk who suffered from anxiety attacks, read Darwin's book and began experimenting with crossing wild and albino mice. His experiments with the sexuality of rodents upset the bishop, and he switched to growing peas in the monastery garden. Mendel worked with plants to determine whether offspring of hybrid peas remained constant over generations. Mendel's discovery, at once both simple and complex, was groundbreaking. He showed that, over generations, hybrids revert to their parents, and lost traits reappear generations later. Known as atavism, the idea contrasted with nineteenth-century views of progressive evolution. Mendel's practical discovery did not quite fit with ideas of continued improvement toward perfection. His findings made it clear that organisms might indeed revert to earlier states of being, meaning throwbacks were quite normal and could not be eliminated. Mendel's laws, as they became known, were the law of

1. H. Cecil Pawson, *Robert Bakewell: Pioneer Livestock Breeder;* Robert C. Olby, *Origins of Mendelism,* 17; George M. Rommel, "Essentials of Animal Breeding," 28–29.
2. Olby, *Origins of Mendelism,* 59, 61, 69; John H. Perkins, *Geopolitics and the Green Revolution: Wheat, Genes, and the Cold War,* 63.

segregation and the law of independent assortment. He provided theory, pointing out that traits are inherited separately and that they may seem to disappear for a generation or two, then reappear again. He believed that traits passed from parent to offspring in units that could be identified, predicted, and sorted mathematically. A radical idea, it laid the foundation for twentieth-century discoveries of genes, chromosomes, and DNA molecules.[3]

Mendel delivered a paper on his results, titled "Experiments in Plant Hybridization," to a local scientific society in 1865, and after its publication in the *Proceedings of the Natural History Society of Brünn*, he sent reprints to about forty scientists in the field. He received no response. Historians believe many of them went unread; some remained uncut, the folded edges of the pages never slit apart. It was not that his work was ignored; rather, the neglect was due more to the lack of development of the field. Eventually, scientific breeding became better understood, and in 1881 Mendel's paper was cited fifteen times in a book about plant science. Unfortunately, after his death in 1884, his successor at the monastery burned all of Mendel's papers.[4]

Thirty years later, Mendel's efforts garnered attention when scientists in separate countries arrived at conclusions similar to Mendel's. Working separately and unaware of each other's results, Hugo de Vries, a Dutch hybridist; Karl Correns, a botany professor at the University of Tübingen; and Erich von Tschermak, a graduate student in Vienna, all published similar papers in the spring of 1900. Tschermak studied the practical breeding of grain crop plants, while the other two focused on plant physiology and evolution. De Vries' work was disseminated to British and American wheat breeders, whereas Tschermak's contact was largely with Swedish wheat scientists.

As the nineteenth century closed, the field evolved enough to begin reexamining and rethinking Mendel's findings. Yet his work needed to be reintroduced to the scientific community. In 1899, the first International Conference on Hybridization and Plant Breeding was held by the Royal Horticultural Society. It was at that conference that William Bateson, of Cambridge University, heard De Vries present his research.

---

3. Robin Marantz Henig, *The Monk in the Garden: The Lost and Found Genius of Gregor Mendel, the Father of Genetics*, 16, 157, 7; Olby, *Origins of Mendelism*, 115, 177.

4. Henig, *Monk in the Garden*, 143, 157. The 1881 publication was *Die Pflanzen-Mischlinge*, by Wilhelm Olbers Focke.

De Vries referred to Mendel's work in a footnote, which caused Bateson to resurrect the long-forgotten 1866 paper and promote it to British scientists. Robin Marantz Henig, Mendel's biographer, believes that "if not for Bateson, Mendel's pea experiments might never have become the unifying starting point of genetics." Interestingly, Bateson did more than present Mendel's work; he took up the persona he attributed to the monk, adopting chess (Mendel's favorite pastime) and cigar smoking (Mendel's habit) and subscribing to a humor magazine he discovered had been Mendel's favorite. Filled with enthusiasm for what Mendel's ideas meant, Bateson announced that the field of biology was about to change. "Presently steam will be introduced into Biology and wooden ships of this class won't sell," he stated, referring to examining plants and animals as they existed. Realizing how valuable the new understanding of inheritance could be, Bateson became the "apostle of Mendelism in England."[5]

The story surrounding Mendel's "forgotten" work and its "rediscovery" has been an ongoing argument among scholars. John H. Perkins believes Mendel's paper can best be understood as an idea that appeared before its time and that its reconsideration in the late 1890s came about not because Mendel's work was rediscovered but because the field had advanced to the point where his work became relevant.[6]

Mendel's work helped explain what had eluded scientists and practical agriculturalists: how species change over time through selective breeding. Hybrids, what European scientists referred to as "bastards," or "sports," remained elusive. Breeders recognized that selective breeding did not cause sports; they were a natural surprise that could not be explained—or avoided. Termed *atavism*, the way inherited characteristics disappear for several generations then inexplicably reappear, the process stumped scientists and agriculturalists alike. Understanding how traits passed on through the generations offered great potential: if the process could be understood and controlled, new varieties, races, or strains might be created from such accidents or mutations.[7]

Whereas the European "rediscoverers" of Mendel's ideas have received wide attention, another researcher, an American at an obscure

---

5. Henig, *Monk in the Garden*, 207, 198; Olby, *Origins of Mendelism*, 132.

6. J. H. Perkins, *Geopolitics*, 55.

7. Liberty Hyde Bailey, *Plant-Breeding, Being Six Lectures upon the Amelioration of Domestic Plants*, 13.

college in the Far West, reached the same conclusions, albeit to lesser fanfare. In 1894, William Jasper Spillman arrived at the Washington State Agricultural College and School of Science. The school, newly opened the previous year, sat on the edge of Pullman, a town of about one thousand residents, in a wheat-growing belt known as the Palouse. Spillman was young, enthusiastic, and popular with students, most of whom were still attending the college's preparatory classes, as the area lacked high schools. In 1892, the college enrolled eighty-five students, sixty-one in the preparatory program. Recruited by college president Enoch Bryan, Spillman accepted a teaching and research position at the fledgling college, agreeing to coach the football team with the use of a handbook. The state agricultural experiment station was part of the college and was described by the local newspaper: "At each station will be an officer of the college, who will have a corps of assistants and a certain number of students. At each station a study of the surrounding locality will be made and experiments will be performed." Work would focus on "the intervention of science." Spillman began his work focusing on forage crops, an essential source of fuel for draft animals. He enthusiastically embarked on a variety of experiments, growing as many as a thousand varieties of grasses, clovers, and specialty plants in the college's "grass garden," near the campus.[8]

Spillman was an early advocate of linking research and extension, of sharing his views and eliciting the views of area farmers. Because the young agricultural scientist believed that forage crops would better suit the surrounding Palouse region, an upland rolling prairie that was devoted almost exclusively to wheat growing, he urged farmers to diversify. He justified his position by explaining to farmers that "the reason for growing one crop or another, aside from moral considerations, is a matter of profit on investment and labor. If the continuous raising of wheat is better for the welfare of those concerned than the growing of forage crops, then wheat should be grown." However, he cautioned that, "while prices are good and the land still abundantly fertile, wheat growing is profitable." He warned about soil depletion: "As time goes on, the average yield of wheat will decrease to a point where wheat culture

8. Bryan, *Second Annual Report: Washington Agricultural College and School of Science,* 7; "College Opening," *Pullman (WA) Herald,* September 1, 1893; "College Regents," *Pullman Herald,* May 9, 1891; E. N. Bressman, "Spillman's Work on Plant Breeding," 273; Spillman, *Forage Plants in Washington* (1900), 4.

will be profitable only in exceptional years." Besides sustainability, he argued that farmers were ignoring local markets: "At the present time people in many a village of the state of Washington are eating butter, eggs, and bacon from Iowa. Beef cattle are not to be had. There has not been a mutton in the local butcher shop for two weeks." Farmers should not rely on only one crop for their security, particularly wheat, which was notoriously unpredictable and in oversupply. "We have reached a point where there is immediate demand for animal products far in excess of the supply. Under such conditions, farmers can make money by raising beef, hogs, chickens, and dairy cows," he advised. Diversified farming, which included livestock, made better economic sense.[9]

Spillman told the farmers that large areas of the region were simply too dry for successful continuous cropping of wheat. He believed that many parts of the region were best suited for grazing. "The wild grasses upon it were originally both abundant and nutritious," he pointed out. But past practices led to problems: "The native grasses have been largely killed out by over-stocking with cattle and sheep." His quest was to determine how to regrass the ranges to support controlled grazing. He sent questionnaires to farmers in eastern Washington and developed a regional guide to forage crops based on rainfall and soil types. He encouraged farmers to use crop rotations that included leguminous plants to restore nitrogen to their soil and to avoid cutting hay year after year, essentially selling off "the fertility of the farm." Grazing allowed animals to manure and maintain the soil.[10]

In 1895, one year after his arrival at the school, experimental work was focused on feeding wheat to hogs, forage crops from worldwide sources, cultivation of native grasses as forage, and field experiments with wheat, oats, barley, and corn. Using wheat seed stock largely from Kansas, he conducted research in the college test gardens that focused on depth of plowing and rate of seeding in an effort to obtain better results from fall sown crops. The results were confusing; little correlation could be made between farming practices and variety of wheat used. The best conclusion to be reached was that early sowing was more successful. Spillman also concluded that the efforts were seriously underfunded, with inadequate labor to complete many of the projects. One

9. Spillman, *Forage Plants in Washington*, 8.
10. Ibid., 9, 27.

series of experiments had to be abandoned due to lack of labor to complete the work. The equipment added that year was limited to four cows, two calves, three pigs, a horse team, a sixty-five-ton silo, a subsoil plow, a grain separator, and a farm engine. The plow and separator, along with ten gallons of carbon disulfide for experiments on killing ground squirrels, were donated by businesses.

An adjunct to the establishment of state experiment stations was the requirement that research be shared with farmers. In January 1895, during Spillman's first year at the college, the agricultural department sponsored the first Winter School for Farmers, a joint undertaking by the college and the experiment station. It lasted three weeks, and 304 farmers attended. The farmers were eager to learn from the experts. "I have never seen a more earnest or enthusiastic body of students. I am convinced more and more that the close alliance between *investigation* and *instruction* is alike of the highest importance to the college, the experiment station and the people," Enoch A. Bryan, director of the Washington State Experiment Station, noted in the year-end report.[11]

Spillman's ideas about diversified sustainable farming may have found a few receptive ears, but farmers in the Palouse region of the inland Northwest preferred growing wheat, a crop that was a speculative venture rather than traditional husbandry. Wheat, in the 1890s, was a global product, sold on an international market. Wheat growing was heavily capitalized, highly risky, yet held enormous potential for profit. Unlike dairying, orchards, or vineyards, growing wheat was a venture that appealed to investors who relocated to the region's small towns, hiring labor or leasing out portions of vast farms. Capitalization for land and machinery was all-important, rather than knowledge of complex agricultural techniques. Viticulture required complicated grafting and trimming; dairying required knowledge about bacteria, fat content, and feeding patterns; even raising chickens commercially required one to have a background in hatchery production. To grow wheat, or other grains, soil preparation and planting were crucial; after that, nature controlled the crop. Palouse farmers did not irrigate or do anything else to the crop once it was planted. To make it even more profitable, farmers often let the grains lost during harvest reseed the fields a second year.

11. Bryan, *Fifth Annual Report: Washington Agricultural Experiment Station, Washington State Department of Agriculture*, 8.

Although the region had long been a successful grazing area, the coming of railroads transformed the region into export monocultural agriculture based on easily grown wheat. Little wheat was sold locally; it was the perfect crop for distant markets, as it was compact, stored easily, and could be sold months, even years, after harvest. It was the perfect crop for late-nineteenth-century capitalists and investors. The wheat trade, centered in Liverpool and Chicago, connected to the farm and country elevator with telegraphs, ticker tape machines, and railroads. With a bevy of unpredictable problems, from international demand to plant diseases and the weather, the international wheat market was a gambler's paradise.

In the years between 1870 and 1900, wheat growing exploded on the northern plains and in California's San Joaquin Valley, Argentina's pampas, British India, and the Palouse region of eastern Washington. Argentina provides one example of how rapid change came about. Argentina first exported wheat in 1871, and by 1900 it was the world's third-largest exporter. There, too, the Ministry of Agriculture recommended diversified farming, but growers ignored the advice and tried to double their profits by doubling their acreage, until the entire pampas seemed to be in wheat production. In the United States, the "huge romantic West" extolled in novels no longer existed at the turn of the century, and much of the reason was due to the speculative nature of industrialized wheat growing. Novelist Frank Norris described the situation at the time in his two novels about the wheat industry: *The Octopus* (1901), which examined farming in the San Joaquin Valley, and *The Pit* (1903), a novel about the Chicago wheat-trading world. Norris revealed what farmers already knew: farmers across the wheat-growing West were faced with new technologies and challenges that did not quite fit the simple yeoman image. Stock tickers kept wheat growers awake nights as they watched the wire for the Chicago, Minneapolis, Duluth, New York, and Liverpool prices. Rail rates meant profit or loss, and the vagaries of weather around the world became crucial factors. Wheat growers participated in a larger competitive world. "They became part of the enormous whole, a unit in the vast agglomeration of wheat land in the whole world round," Norris wrote, "feeling the effects of causes thousands of miles distant—a drought on the prairies of Dakota, a rain on the plains of India, a frost on the Russian steppes, a hot wind on the llanos of the Argentine." Farmers across the wheat-growing regions faced the dilemma of growing a crop for an unpredictable, often saturated, world market.[12]

Successful wheat growing required three elements: a lot of land, a favorable climate, and access to rail lines. Western U.S. farmers had plenty of rail lines; in fact, the railroads had recruited farmers to settle along their routes in order to create customers for rail shipment. Palouse farmers had vast tracts of tillable land because the region was naturally a fertile grassland with deep topsoil. The climate, however, was unpredictable. Crops had to be seeded while the soil was still moist in the spring, but early enough to get in an adequate growing season before autumn rains and frost arrived. When the heads filled for harvest, thousands of acres had to be cut and threshed in a few weeks, before the weather ruined the grain.

Timing, machinery, and cheap seasonal labor were essential to success, even if prices, markets, and freight rates were amenable. In *The Octopus*, which examined the struggle of farmers against powerful industrial and transportation forces arrayed against them, Norris painted a picture of late-1890s wheat farming:

> Everything seemed to combine to lower the price of wheat. The extension of wheat areas always exceeded increase of population; competition was growing fiercer every year. The farmer's profits were the object of attack from a score of different quarters. It was a flock of vultures descending upon a common prey—the commission merchant, the elevator combine, the mixing-house ring, the banks, the warehouse men, the labouring man, and above all, the railroad. Steadily the Liverpool buyers cut and cut and cut. Everything, every element of the world's markets, tended to force down the price to the lowest possible figure at which it could be profitably farmed.[13]

Though wild profits could be made in the trading pits, profit margins were usually extremely narrow for the farmer. Without adequate

12. James R. Scobie, *Revolution on the Pampas: A Social History of Argentine Wheat, 1860–1910*, 70; Norris, *The Octopus: A Story of California*, 34.
13. Norris, *Octopus*, 35.

storage facilities, the crop often sat beside a rail station or piled near a shipping facility. Because farmers could not store their crops, and often had liens against them due at harvest, everything was sold at harvest time. Without quality standards for grain, buyers would often deduct for dirt or mold in the grain, then sell it at a premium to the flour mills.

The year 1894 was difficult for the U.S. wheat market, as "there are speculators who believe that those good old days when $1 a bushel was the rule will never come again, and that 75 cents may be looked to as the future top-notch quotation," the Pullman newspaper reported. Wheat was selling at less than the cost of production, with sixty cents per bushel figured to be the cost of production east of the Mississippi and higher costs due to transportation to market for western growers. Palouse farmers were isolated from the distant markets of Europe and were competing with increasing acreages in Argentina, India, and Russia. Palouse businessmen, bankers, and farmers had no answers. They turned to science, in the form of the college experiment station.[14]

Spillman found himself in the center of this highly competitive, highly speculative industry. He had grown up on a small Missouri farm, cutting grain with a scythe and bundling it by hand, taking a herd of cattle to St. Louis as a way to earn passage to college. The farm boy with a good education realized that the region's farmers were pressing for scientific information, but growing hay did not interest them like wheat. They wanted an edge that would help them succeed at the only crop that offered them a chance to be spectacularly successful. Spillman began experimenting with wheat in his "grass garden" at the agricultural college. He sent questionnaires to the region's wheat growers, asking about the seed varieties they used and entering those data in his records about soil and rainfall. Nearly 150 growers responded, revealing that the chief varieties grown were Little Club, Bluestem, and Red Chaff. Little Club grew in wet areas, Red Chaff in moderate regions, and Bluestem in the driest. They were all wheats grown in warmer climates, such as California and Oregon, where they were planted in the spring and harvested that autumn. The farmers in eastern Washington, however, with a shorter summer growing season, were planting them in the fall to extend the growing time and gain larger yields of up to 21 percent. The winter weather often wiped out the crop, though. Because the yields

14. "The Future of Wheat," *Pullman Herald*, March 23, 1894.

were so much greater when luck favored them with a mild winter, the wheat growers—inveterate gamblers—hoped Spillman could figure out a way to better their odds.[15]

In 1899, Spillman and his students divided sections, one-sixtieth of an acre in size, and seeded all the wheat varieties they could get their hands on. Each seed was charted, pedigreed, and planted and the results noted and assessed. A friend later explained that about this time, "Dr. Spillman was sunburned from being in the field so much studying hybrid plants and had to go to the hospital."[16]

Spillman decided to try crossing varieties, particularly one called Turkey *(Triticum vulgare)*, which was very winter hardy, and Little Club *(Triticum compactum)*, which had strong straw (which would not bend and flatten on the ground during windstorms) and tight heads that did not scatter the seeds while harvesting. He and the students made fourteen crosses by hand-pollinating the plants and obtained 215 plants, of which 149 were hybrids, the rest identical to the female plant. When mature, the wheat heads were sorted according to characteristics, revealing that when one parent had long bearded heads and the other short beardless heads, the resulting offspring could be divided into six types: two with long heads like one parent, two with short heads like the other, and two that were intermediate. Of those three groups, one each had beards, while the others did not. In some cases, one parent plant had velvet chaff and no beards, which resulted in six types where the velvet chaff replaced the beards. Spillman created tables to show the results, particularly those of the third generation, when some characteristics reverted to the parent plant, while others did not.

No variations occurred in the first generation, but "when the heads appeared on the second generation, a remarkable state of affairs was seen to exist," Spillman explained. "At the first glance it appeared that each of the hybrids had split up into all sorts of types. But closer inspection showed that in every case but one ... the forms in each plat were simply combinations of the characters of the parent form." This was surprising, because Spillman, like most animal and plant breeders, held the same views as Darwin: that traits from parents blended over the

15. Stephen S. Jones and Molly M. Cadle, "Spillman, Gaines, and Vogel—Building a Foundation," 26; W. T. Yamazaki and C. T. Greenwood, eds., *Soft Wheat: Production, Breeding, Milling, and Uses,* 68.
16. Bressman, "Spillman's Work," 274.

generations to create new forms. Here was evidence that those traits separated and came together, never "blending" at all. After sorting, classifying, and correlating the results, Spillman observed that after the crosses (more than 1,000 were made), results became predictable. "This suggested the idea that perhaps a hybrid tended to produce certain definite types, and possibly in definite proportions." He realized that "it seems possible to predict, in the main, what types will result from crossing any two established varieties, and approximately the proportion of each type that will appear in the second generation." His ability and training in mathematics helped him construct a formula to figure out that there was a predictable pattern to the inheritance of characteristics.[17]

A teachers' association met at the college in 1901, and Spillman spoke to them about his wheat research, telling them about the segregation and recombination of characters that had appeared in the parent varieties. Later that year, he read a paper on his findings to a meeting of the Association of American Agricultural Colleges and Experimental Stations, in Washington, D.C. He took along a large piece of cloth, on which he had glued wheat samples, showing how the various traits continued over generations of plant crosses. Spillman showed clearly that traits combined and recombined, rather than blending together, as most scientists had believed. His work, done in the West, replicated the same findings that the papers of De Vries, Tschermak, and Correns, the European Mendelian "rediscoverers," had published earlier that year.[18]

The Department of Agriculture found Spillman's work of interest; Willett M. Hays, an agriculture professor at the University of Minnesota who had organized the meeting (and who would soon become assistant secretary of agriculture), was said to have been "in a front row seat and much surprised at the results shown." Hays had been selectively breeding wheat for some time but had not understood the pattern of inheritance that Spillman—and Mendel—explained. Spillman received a job offer for the position of agrostologist at the Department of Agriculture, which he accepted, resigning his position at Washington State on

17. Spillman, quoted in *Experiments in Genetics*, by Charles Chamberlain Hurst, 110; Jones and Cadle, "Spillman, Gaines, and Vogel," 27; Spillman, "The Hybrid Wheats" (1904); Bryan, *Fifth Annual Report*, 7–8; Bryan, *Sixth Annual Report: Washington Agricultural Experiment Station, Washington State Department of Agriculture*, 5–9.

18. Bressman, "Spillman's Work," 274.

December 31, 1901. He wrote to the college's president, Enoch Bryan: "The work has been of a character to inspire supreme effort, because of its opportunities for usefulness." His tenure at the school had been "the pleasantest years of my life."[19]

His successor at the college, E. E. Elliott, continued the hybrid wheat investigations for a few years, and as director of the Washington State Experiment Station, he noted in his 1902 report that "there is reason to believe that some important laws underlying the whole theory of plant breeding have been discovered . . . while at the same time the practical purpose of the experiments will not be lost sight of."[20]

After Spillman's departure, other colleagues continued the wheat-breeding program, selecting the desired seeds and discarding those not winter hardy, and by 1907 six varieties were available to growers. Hybrid 128 became the most popular wheat in the Pacific Northwest, and in 1908 it was seeded on more than forty thousand acres. By 1911, Spillman's varieties were planted on half a million acres.[21]

Spillman's paper "Quantitative Studies on the Transmission of Parental Characters to Hybrid Offspring" was included in the publication of the proceedings of the meeting in 1902 and reprinted in the *Journal of the Royal Horticultural Society* in 1903. British scientists were quick to realize that Spillman had been another of the "codiscoverers" of Mendel's principles. In London, Charles Chamberlain Hurst devoted a chapter in his book *Experiments in Genetics* (1925) to Spillman's wheat studies, noting, "Prof. Spillman's paper is of great biological and practical importance." Lauding Spillman's procedures, he stated: "The experiments, on the whole, seem to have been admirably designed and carefully carried out, especially having regard to the large numbers used; the examination and classification of many thousands of individual characters must have entailed a vast amount of labour and care, for which we are duly grateful to Prof. Spillman and his associates." He remarked that it was "fitting that this important paper should be published in the

---

19. Alfred Charles True, *A History of Agricultural Experimentation and Research in the United States, 1607–1925,* 188; Elliott, *Twelfth Annual Report of the Director of the Washington Agricultural Experiment Station,* 7; Spillman to Bryan, December 9, 1901, cage 250, box 1, file 2, William J. Spillman Papers, Manuscripts, Archives, and Special Collections, Holland Library, Washington State University, Pullman (hereafter cited as MASC).
20. Elliott, *Twelfth Annual Report,* 8.
21. Jones and Cadle, "Spillman, Gaines, and Vogel," 28.

same *Journal* that first published an English translation of Mendel's original paper." For Hurst, it was important that Spillman's work, so well documented and full of data, actually proved that Mendel was correct. It was not so much that Spillman had "rediscovered" Mendel's principles, but that his work, done without knowledge of those ideas, was clearly an "independent confirmation of Mendel's Principles." Hurst passed a copy of Spillman's paper on to Francis Galton, who thanked him for the "important memoir on Wheat Hybrids." "I have studied and shall re-study it carefully," he wrote. Galton's cousin Charles Darwin had passed away by that time, or perhaps he, too, would have read Spillman's wheat findings.[22]

At the Department of Agriculture

In 1902, Spillman arrived at the Department of Agriculture, the administrative center of federal agricultural efforts, as part of a dynamic cadre of select individuals. The department was a hotbed of enthusiastic handpicked young men, eager to put their educations and ambitions to work for the public interest. The scope and staff of the department were expanding quickly, with a focus on scientific research to improve production and efficiency. The staff grew from 2,444 in 1897 to 13,858 by 1912. The budget soared from $3.6 million to $21.1 million. Secretary of Agriculture James Wilson, previously a professor at Iowa Agricultural College, presided over four bureaus: plant industry, soils, chemistry, and forestry. Yet, as USDA historian Wayne Rasmussen pointed out, the early department was not really organized around projects or offices— rather, it was organized around people. Rasmussen described the staff as "brilliant, innovative, and sometimes irascible leaders," and Secretary Wilson largely encouraged their creativity and free expression. In his report that year, Wilson noted that "the policy has been to encourage the most advanced work by placing the responsibility for different lines of investigation and research directly upon the men themselves. It is believed that the best results can be obtained always by assigning men to different lines of work and making them feel the responsibility for its

22. Hurst, *Experiments in Genetics*, 109, refers to it as appearing in *J. Royal Horticultural Society* 27 (1903): 876–93; Hurst, *Experiments in Genetics*, 109, 132.

advancement." Employees were encouraged to "advance in all directions" and to "make their work as thoroughly practical as possible" while remaining grounded in "sound scientific principles." In Spillman's division, the Bureau of Plant Industry, "there is an earnest corps of workers, each knowing his duty and performing it with all the energy at his command." Wilson handled the variety of egos and temperaments well, as he juggled a variety of projects and personalities, such as Harvey Wiley's incessant promotion of his chemists in the Bureau of Chemistry with his "Poison Squad" of food samplers and national forester Gifford Pinchot's band of enthusiastic forest rangers. But Wilson was an experienced politician, already having served three terms in the Iowa legislature and three terms in the U.S. House, something that stood him in good stead in both his department and Washington.[23]

The 1890s had been a violent era, filled with labor strikes, financial depression, and unrest, both urban and rural. Whereas labor strikes and riots in cities had garnered attention and required federal law enforcement to keep order, a rural movement to create a third political party, the Populists, was more problematic. The People's Party gained strength as regional and class factionalism split the nation. At a time when half the nation's population still resided on farms, agrarian unrest threatened to create national problems.

The progressive movement, initiated by educated middle-class reformers, evolved as an urban response to the turbulent events, including disorder in the countryside, according to historian Robert H. Wiebe. Centralization became the hallmark for both corporations and government, as the nation adopted administrative centralization as a response to localism.[24]

The Department of Agriculture, through education and service, could allay some of that rural discontent. Given cabinet status in 1889, the department promised to become an antidote to "the growing dissatisfaction among the rural half of the nation, evident in the surge of organizations such as the Farmers' Educational and Cooperative Union

23. Wayne D. Rasmussen and Gladys L. Baker, *The Department of Agriculture*, 14, 15; Secretary of Agriculture, *Annual Reports of the Department of Agriculture, Fiscal Year Ended June 30, 1902*, 49–50, xv.

24. See Wiebe, *The Search for Order, 1877–1920*; and David B. Danbom, *The Resisted Revolution: Urban America and the Industrialization of Agriculture, 1900–1930*.

and the American Society of Equity," USDA historian Wayne Rasmussen noted. Half of the nation's population resided on farms, and farmers needed professional assistance. "The education of producers from the field, so long neglected, has recently been undertaken in earnest in the United States," Secretary Wilson explained. Export agriculture was most important, and through it the department would justify its budget: "The future will still further show the value of science applied to the farm." Not only would scientific agriculture boost national export income, but it would also solve political problems that had emerged in the South and West. In those regions, where wheat and cotton prices fluctuated uncontrollably, Grangers and Populists emerged following economic downturns. Offering them technical expertise as a solution to their economic problems might quell dissent. As historian Samuel P. Hays explained, "Markets, machinery, and science, then, transformed American agriculture from a relatively simple operation, requiring little capital and less knowledge into a highly complex affair, demanding increasing amounts of investment, equipment, scientific information, and close attention to markets. The farmer was now irrevocably entwined in the complex industrial system." Hard work was no longer enough; "only as a calculating, alert, and informed businessman, could he survive."[25]

One new field—scientific breeding—promised economic benefits with little cost. Enthusiasm for the field had grown, but it was still in its infancy. Not until 1906 would the term *genetics* be coined at a London conference, and in 1909 the term *genes* would be adopted in Denmark. In 1911, the Punnett square was developed, the rubric construction that simplified the understanding of genetics, making distribution of characteristics easier to understand. In 1903, a group of scientists and government officials gathered to found the American Breeders' Association (ABA) to bring scientists, government officials, and practical breeders together to pursue discoveries. The first meeting was held in a St. Louis high school, organized by the American Association of Agricultural Colleges and Experiment Stations. The gathering of forty to fifty "animal men and plant men" shared their research and theories, in a field newly established due to increased attention to Mendel's work. Intense

25. Rasmussen and Baker, *The Department of Agriculture,* 14; Secretary of Agriculture, *Annual Reports,* cxxiv; Hays, *The Response to Industrialism: 1885–1914,* 15.

interest focused on breeding plants and animals for economic reasons: agricultural production had languished, the urban population had expanded due to industrialism, and a foreign export market beckoned.[26]

Industrialism shaped all facets of American thinking, even seeing animals as production "machines." David Fairchild, a research botanist at the USDA, stated, "Plant and animal breeding is becoming a manufacturing process." Raw materials included domesticated animals and plants, with an infusion of exotic species procured from regions around the globe.[27]

The ABA focused on using genetics for economic gain, urging biologists to turn to what they called "the needs of artificial evolution" for the "common good of the country and the world." The scientific field of breeding management held promise that new strains of plants or breeds of animals could be created that would better suit problematic environments in particular. But they became so mired in arguments over Darwin's ideas that they had "allowed the great economic problems of evolution guided by man to remain almost a virgin field," one speaker noted. Economic interests remained central, attendees realized, despite the new appeal of Mendel's work to theorists. "Many of these [researchers] have hardly grasped the vast economic interests which are at stake," one speaker reminded. Recognizing that hybrids presented an opportunity to create new forms of plants or animals, figuring out how to understand—and control—them captured the field. If a superior hybrid could be made to reproduce, then a superior (or at least different) breed of plant or animal might be obtained. Potential existed in the "Shakespeares in every species," according to one scientist, and the ABA must hunt for them, "and, having secured their blood, multiply it until the blood of the old species or old variety is supplanted by the more valuable blood of the occasional plant or animal."[28]

Finding those singular examples of perfection presented a challenge, along with understanding how to control them to create new lines of individuals. New creations, however, must be practical, not exotic. Pragmatic goals pervaded the ABA efforts.[29]

26. Henig, *Monk in the Garden*, 228, 239, 215.
27. American Breeders' Association, *Proceedings*, 1:92; Fairchild, *The World Was My Garden: Travels of a Plant Explorer.*
28. American Breeders' Association, *Proceedings*, 1:11.
29. Ibid., 9, 12.

Presenters discussed a variety of experimentation occurring across the country: attempts to create a fertile race of mules that ate lightly like donkeys but pulled with the strength of draft horses seemed most economically desirable. The mules, however, remained infertile, a problem researchers could not overcome. Animal breeders attempted other hybrids of domestic animals and reached out to wild species for fresh bloodlines as well. Cattelo, a hybrid mixed from breeds of cattle and buffalo, and sometimes yaks, were also sterile. Cattle breeding did yield positive results; mixing imported African Zebu cattle with British breeds created the Brahmin breed in Texas, which better suited the hot climate there.[30]

Spillman, an active participant in the ABA during its formative years, explained how Mendelian inheritance patterns affected the problem of recessive traits, those that showed up unexpectedly after one generation. Because atavism was usually considered a defect (few "Shakespeares" appeared in any species), managed breeding appeared to be one way to ameliorate its appearance in future generations. Spillman explained how particular traits allowed one to breed cattle without horns. By breeding animals that had horns but carried a recessive trait for hornlessness, a herd of hornless, or polled, cattle could be obtained. Spillman, however, also presented one of the few ethical questions the group considered. He noted that although many people wanted polled cattle, "it is not for me to say whether or not it is desirable to remove the horns from any breed of cattle." He realized the power of the concepts under discussion: genetic manipulation was too easy. He pointed out that "the same may be done with reference to any character in which a change may be desired just as soon as we have learned what are the unit characters involved." Whether decisions about characters and their propagation (or elimination) would be made wisely was another matter that the group did not pursue. That moral question, whether man should tamper with nature and on what terms and why, did not arise again in the conferences. Manipulating plants and animals for increased yields and profits reflected progressive ideals of control and management for the common good.[31]

30. Ibid., 14. For cattelo, see American Breeders' Association, *Proceedings*, 6:133; and David C. Rife, *Hybrids*, 46.

31. Spillman, "Mendel's Law in Relation to Animal Breeding" (1905), 176. The movement to poll Herefords and Shorthorns began at about this time and may have

Even though only a few successes could be claimed, the idea that animals resembled efficiently operating machines held great appeal to the ABA members. In 1905, Willet Hays, assistant secretary of agriculture, stated that "as science, inventive genius, constructive skill, business organization, and great market demands at home and abroad have pushed forward things mechanical, so should ways be found of improving these living things which serve as machines for transforming the substance of soil and air and the force of the sun's rays into valuable commodities." Hays compared the complexity of inheritance to mechanical invention; breeding was "more abstruse and more profound than the mechanism of the mightiest locomotive." Yet the promise of harnessing the uniqueness of a single individual continued to prove irresistible. Hays reiterated that the potential for "a single Shakespeare-like individual to be multiplied a million fold as a valuable new variety or breed" held fantastic promise, not unlike sixteenth-century visions of the Cities of Cibola or Fountain of Youth. It was a vision based on mechanical efficiency: "As one machine is more efficient than another, so the blood of one generative cell, or of a small group of generative cells combined into an efficient varietal or breed unit, is more valuable than another."[32]

The Committee on Eugenics

In 1906, the ABA added the Committee on Eugenics, sending the group in a direction far different from its original goals. Eugenics, a term coined by Francis Galton meaning the study of human breeding, grew increasingly popular across the country and influential within the ABA. The committee investigated human heredity with emphasis on "the value of superior blood and the menace to society of inferior blood" and to "suggest methods of improving the heredity of the family, the people, or the race." Charles Davenport, a zoologist formerly at the newly founded University of Chicago, built a eugenic laboratory at Cold Springs Harbor, New York, with support from the Carnegie Foundation. By 1909,

---

been sparked by Spillman's simple explanation. See Everett J. Warwick, "New Breeds and Types."

32. Willett M. Hays, "Breeding Problems," 197.

the goals and projects of the ABA had split, with heavy emphasis on eugenics, which took a negative approach by targeting individual characteristics and advocating sterilization to prevent inheritance.[33]

Inheritance and dysfunctional families were mythologized by the work done on the Jukes family of rural New York State, a clan said to represent how genetic-related traits caused deficiencies, thereby creating for taxpayers the huge financial burdens of hospitals, poorhouses, and jails. New research reveals that the Jukes study was inherently flawed; it was a powerful example of how "scientists distorted research results for ideological and political reasons," according to Scott Christianson. The Jukes study was used to bolster the pseudoscientific theories of eugenicists such as Davenport. The original research was actually a composite of several families, and only 540 of the 709 subjects were actually related by blood. At Cold Springs Harbor, however, researchers added another 2,111 "Jukeses" to the study in 1915, pronouncing the family as an example of inherited immorality, costing the public plenty, and bolstering the arguments for attention to reproductive freedom.[34]

Spillman found himself "too busy" at the USDA to devote time to the group. He did continue in genetics, however, serving as editor of the heredity section for the publication *American Naturalist* from 1908 until 1915, years during which the publication was at the forefront of genetic research in the nation. Genetics had no journal of its own until *Genetics* began publication in 1916.

In 1908, the *American Naturalist* shifted from focusing on paleontology and natural history to evolution and genetics. The American Society of Naturalists declared genetics as the "field of study most likely to unite biologists of all kinds." While Spillman held the reins at the magazine's genetic section the topic received plenty of attention. It was a highly controversial subject, and the open-forum style of the *American Naturalist* allowed scientists to engage in arguments that would have

33. American Breeders' Association, *Proceedings*, 2:11. See also Barbara A. Kimmelman, "The American Breeders' Association: Genetics and Eugenics in an Agricultural Context, 1903–13." For examples of contemporary eugenic attitudes, see S. J. Holmes, *Human Genetics and Its Social Import;* and Thurman B. Rice, *Racial Hygiene: A Practical Discussion of Eugenics and Race Culture.*

34. Christianson, "Bad Seed or Bad Science? The Story of the Notorious Jukes Family," *New York Times*, February 8, 2003, sec. B.

been impossible otherwise. Those arguments were essential to developing the field. Spillman, as editor, played an important role. His tenure there coincided with the journal's most significant period. "The most fruitful period of the *Naturalist* was that following its adoption in 1908 of evolution as the center of its interest," explained L. C. Dunn, marking the journal's one hundredth anniversary in 1966. During Spillman's time there, a sea change occurred in biology as experimental genetics became the new frontier.[35]

The field split along ideological lines as eugenic enthusiasts pushed research in a different direction. The split was rooted in opposing views of Mendelian genetics and the "presence-and-absence" hypothesis. The presence-and-absence hypothesis had been accepted by genetic pioneers including Spillman, Correns, Bateson, and Punnett. The idea revolved around seeing unit characteristics as either present or absent in offspring and not delineating presence or absence of a characteristic as positive or negative. Charles R. Davenport, the eugenic leader, held a different view that contrasted with the presence-and-absence theory. He considered "presence *of* absence" to be a rare phenomenon. Instead, he said that "dominance in heredity appears when a stronger determiner meets a weaker determiner in the germ." For Davenport, the ultimate battle was between strong and weak characteristics. The winner would be evident in the next generation.[36]

As an editor at the *American Naturalist,* Spillman responded to an article about the inheritance of left-handedness by another author, saying there might be environmental factors at play. Children might be trained to right-handedness, an unknown variable that complicated the simplistic idea of Mendelian trait dominance. Spillman, as well as other researchers, tried to move beyond simple unit characters to explain how inheritance worked. He pushed the idea even further in 1910, with an article in the *American Naturalist* that introduced his "teleone" hypothesis: that a chemical substance or enzyme was involved in the transmission of "cell organs having functions which determine character or which influence development." He saw that inheritance involved

35. Dunn, "The *American Naturalist* in American Biology," 487.
36. Davenport, "Determination of Dominance in Mendelian Inheritance," 62. See also George Harrison Shull, "The 'Presence and Absence' Hypothesis," 411.

more than presence or absence of unit characters; rather, it was due to some cellular organ that could influence development, which he dubbed a "teleone." He thought teleones were present in the egg and were inherited from the previous generation. Whether this was a precursor to the idea of finding DNA in cells is hard to say, but genetic research was indeed pushing the limits of available knowledge in 1910.[37]

Eugenics, like agricultural breeding, promised economic benefits, too. "Unfit" individuals were a drain on society, causing public expenditure for poor asylums, orphanages, schools for the feebleminded, reform schools, prisons, and hospitals. Eliminating unfit individuals before they were conceived would benefit all. Examples of backwoods families who multiplied generations of "unfit" individuals became particularly useful in explaining the situation.[38]

The "Country Boy versus the City Boy" Argument

Ironically, as the ideas behind eugenics expanded, critics began to view farm families through a different lens. Were rural poverty and general "backwardness" something farmers perhaps inherited? Could inheritance explain why rural communities were falling behind as the urban centers grew? William Spillman, in an article for *Science* magazine in 1909, "Education and the Trades," included the remark that country children, despite limited school facilities, were superior applied thinkers in comparison with urban children. If given a practical problem, farm children were more likely to find the solution than city children, he claimed.

His assertions set off a series of back-and-forth responses between Spillman and Frederick Adams Woods, a researcher who was studying inheritance in royal families. Woods argued that Spillman's assertion clashed with current research and used the *Who's Who* to show that a survey of fifty-year-old leaders in the United States revealed that a

37. Spillman, "A Method of Calculating the Percentage of Recessives from Incomplete Data" (1910), 383, 384; Francis Ramaley, "Mendelian Proportions and the Increase of Recessives"; Spillman, "Mendelian Phenomena without De Vriesian Theory" (1910).

38. Rice, *Racial Hygiene*, 107; Clifton F. Hodge and Jean Dawson, *Civic Biology: A Textbook of Problems, Local and National, That Can Be Solved Only by Civic Cooperation*, 17.

larger proportion of those listed had been born in urban than in rural areas. "Talent," he wrote, "tends to be drawn by, and to locate itself in the great centers of human activity." And, once in the city, people had children who would necessarily inherit their superior talent. The *Who's Who* list supported his theory. Environmental effects could not be measured, but "mental heredity" could, he added.

Spillman responded by addressing the question of environment, noting that Adams's study of royalty was based on a uniform environment, "the best possible for the development of character and ability." He challenged Woods to study "some class of human beings subjected to an unfavorable environment." Spillman believed Woods would then find that "even in that [royal] class, native ability and natural impulses would prove to be purely a matter of heredity; but that character and actual ability would be found to be profoundly modified by environment." Spillman believed that, "in fact, the whole experience of the human race speaks for this assumption." If Woods was correct, Spillman pointed out, "why should the state go to the expense of maintaining schools, for a man's effectiveness would not depend on his environment but upon his inheritance?"[39]

Undaunted, Woods continued to argue for "nature over nurture" in several articles that became progressively longer and more densely filled with arguments. In one, he included figures to show that Massachusetts produced twenty-nine times as many scientists as eight southern states, that "millions of Negroes have been to school and yet one would scarcely know where to find a single example of a Negro scientist," and that "with the exception of Virginia the entire country to the south of New York has done almost nothing in producing our greatest Americans." Last, he added, "As for the poor whites of the south, they are certainly not the stock from which one would expect scientists."[40]

Spillman replied in his article "The Country Boy" that although talented individuals may indeed seek the cities, the urban advantage was "due to the fact that the cities in general have better school facilities than the country." Success came from educational opportunity, he argued.

39. Woods, "City Boys versus Country Boys," 578; Woods, "The Birthplaces of Leading Americans and the Question of Heredity," 19.

40. Woods, "American Men of Science and the Question of Heredity," 205–8.

He had gathered statistics of his own to match Woods's, arguing that presidents, governors, cabinet officers, and even railway presidents were more than 50 percent from rural areas. He excerpted statements from business leaders extolling the country life for independence, achievement, and character. Woods continued the argument into 1914; by then, he was a faculty member at the Massachusetts Institute of Technology.[41]

The "country boy versus the city boy" argument was part of a much larger movement exemplified in 1908 by President Theodore Roosevelt's Country Life Commission, a presidential advisory panel of sorts that attempted to bring rural people up to the nation's rising standards. Bringing country people into a consumer economy seemed right for the times, particularly when the Country Life Commission was well funded by John D. Rockefeller's General Education Board (GEB), a virtual philanthropic octopus. Accepting the idea that "talent" had already deserted the countryside, the Country Life Commission would work on creating social and cultural change in rural America.

Genetics, meanwhile, which held so much promise earlier, was adrift as theoreticians found economic applications elusive. In 1912, Spillman, president of the Washington Botanical Society, addressed the group, explaining that the problem with the field was due not to lack of experimentation but to a lack of theory. "We need someone to put meaning into these facts. We are in the position of a man lost in a wilderness. What he needs to find is a road." There were plenty of researchers with ample facts, but how to interpret the findings in a conclusive manner had eluded the field of genetics. "Just at present the supply of theories is almost exhausted," Spillman told his fellow botanists. "At present Mendelists are plodding along practically without working theories." By 1920, plant breeders no longer referred to Darwin, De Vries, or evolution—they were no longer seen as critical. The idea of creating unique entities no longer excited the field; the work became simply to select and cross parental varieties to replicate useful characteristics. The idea of finding a "Shakespeare in every species" had faded.[42]

41. Spillman, "The Country Boy" (1909), 405; Woods, "Sovereigns and the Supposed Influence of Opportunity," 905.

42. J. H. Perkins, *Geopolitics*, 64; Spillman, "The Present Status of the Genetics Problem" (1912), 766.

The Wheat Market

Spillman's departure from the Washington State Agricultural College and School of Science occurred as the region spun into large-scale wheat farming, even as competition for export markets increased abroad. California's wheat industry had fallen from twenty-eight million bushels in 1900 to six million by 1912, due to soil exhaustion and a switch to other crops. Wheat production soared in the Palouse country of eastern Washington, with huge harvests on expansive farms. Washington had some of the largest wheat farms in the world. In 1907, R. C. Mc-Croskey, farming at Garfield, Washington, harvested one field a thousand acres in size. Land was at a premium in the Palouse, where eight hundred–acre wheat farms sold for three hundred dollars an acre. Buyers fought for the wheat in order to speculate with it on a world market, as hopes of wars in far-off places like Russia, Japan, and South Africa tantalized farmers with potential markets. The wheat experiments at the local college had paid off handsomely: "Spillman's efforts led to several high yielding, shatter-resistant, stiff-strawed, winter-hardy cultivars." The improved wheat seed boosted production and made machine harvesting more successful. By 1907, the region's wheat crop was huge, and most of it was marketed elsewhere.[43]

Wheat breeders like Spillman, and the others at agricultural experiment stations scattered across the country, were basic to the revolution in farming. Cheap bread, necessary to keep wages for industrial workers low, created a market for flour in industrializing nations. As wheat had become an international commodity in the mid-nineteenth century, following Britain's repeal of the Corn Laws, farmers in temperate cereal-producing regions had to figure out how to produce more wheat cheaper than other farmers in order to survive. "Higher yields at lower costs were key to economic survival," explained historian and biologist John Perkins. "Darwin and Mendel brought a sense of order, prediction, and manipulation into the study of variation. These perceptions were critical to the construction of an agricultural plant science that served as a base for a complete industrialization and commercialization of agriculture."

43. A. M. Ten Eyck, *Wheat*, 123. "Washington Items," *Coeur d'Alene (ID) Independent,* July 28, 1894; Yamazaki and Greenwood, *Soft Wheat,* 60.

Plant breeders allowed farmers to increase their yields while cutting back on human labor inputs, as capital replaced people on twentieth-century farms. Wheats like Spillman's hybrids suited industrialized farming. Stronger stems and heads that did not shatter worked well with mechanical harvesters. Later wheat innovations focused on plants that did well with pesticides and fertilizers.[44]

As science lowered the cost of production, it also fostered an exodus of unnecessary labor off U.S. farms. Hybrid plant breeders, through their efforts to increase profitability with higher yields, were one of the major influences leading to the depopulation of rural America. As yields increased, commodity prices fell; farmers expanded their acreage to make up the difference, and larger farms, needing less labor, pushed people to urban jobs in industry and services. That exodus came to the attention of the Department of Agriculture in the early 1900s. Certainly, the department encompassed forestlands, retail food and drug protection, even a road department. But central to its very existence were farmers. How could they be kept on the farm while expanding production? That question was to consume the rest of William Spillman's professional career, as he sought to find a balance between the advantages of science and its effects on rural America. To do so, he shifted from genetics to farm management and outreach programs, working to establish the science of farm management and the Agricultural Extension Service.[45]

44. J. H. Perkins, *Geopolitics*, 46, 44.
45. Ibid., 15.

2

Taking Scientific Agriculture to the Public

A skilled speaker, enthusiastic teacher, and innovative scientist, Spillman wanted to deliver the new scientific knowledge emerging in the Progressive Era to wide audiences. His vision was not easily or quickly realized. How to bring together widely diverse interest groups took some missteps, a lot of work, and an element of politics. Farmers, educators, and business interests all wanted input and control of programs. Spillman, receptive to new ideas as well as flexible in approach when necessary to attain a goal, became a key facilitator in linking government, state agricultural colleges, and farmers nationwide.

The going was difficult because public agricultural education had no firm goals or direction; indeed, sponsors varied widely from political groups to business interests. There was not even agreement on a common audience. Some agricultural educators thought "backward" farmers needed education to rise from poverty; others wanted to work with already successful farmers, those most likely to grasp complex new ideas. Ultimately, Spillman crafted what became a very successful hybrid program: college-trained agents worked under the direction of the federal government, using information gained at state experiment stations, to assist farmers with local programs.

Integral to formalized agricultural education, Congress created a nationwide system of publicly funded universities, based on the Morrill Land Grant and College Acts of 1862 and 1890, which provided federal funding for a national system of higher education focused on agriculture. The 1862 Morrill Act authorized states to establish agricultural colleges and endowed them with grants of public land. In 1887, the Hatch Act funded agricultural experiment stations located at agricultural colleges.

In 1890, the second Morrill Act provided additional funds for agricultural and mechanical colleges. By 1896, there were twenty-five thousand students enrolled in land-grant colleges. In 1914, the Smith-Lever Act would solidify federal support for agricultural education delivered by extension agents connected with state colleges. The network of legislation was in many cases a compromise among stakeholders: farmers, educators, bureaucrats, and industry all shaped how America's system of agricultural education would be administered and delivered.[1]

State colleges, including land-grant institutions freshly built in the West, met resistance because they were political entities, often established in areas where local residents had no inclination to support such an institution. Farmers criticized the political dynamic that founded and supported state colleges, but opposition was not limited to rural folk. In 1891, a critical article in the *Atlantic Monthly* pointed out "the conviction is deepening that the founding of a college is not necessarily a blessing to the community." State colleges were governed by boards of regents selected by the governor and dependent on the legislature for support, which created a very politically dependent institution. Because a state college necessarily was "a political body," that was its strength at times but also its weakness. In the Gilded Age, the public viewed state colleges as too political, and expansion was "retarded by a lack of public sympathy." However, "the public schools and public colleges had united," according to the *Atlantic Monthly*. "In consequence, the latter is already taking deeper hold on the affections of the people," which was "likely to be furthered by the movement for university extension, already promising so well in Wisconsin." The public's support for education began at the bottom with local public schools and expanded to include state colleges and normal schools. Land-grant colleges, established by the federal government, had to find their own way to gain public confidence.[2]

Roy Scott, a leading scholar in the history of agricultural education, explains that resistance to early land-grant colleges was due to a bevy of reasons: there was actually little scientific knowledge to pass on to farmers, the faculty was often incompetent, classicists held agriculture

---

1. Louis Ferleger, "Arming American Agriculture for the Twentieth Century: How the USDA's Top Managers Promoted Agricultural Development," 213; Lawrence A. Cremin, *The Transformation of the School: Progressivism in American Education, 1876–1957.*
2. George E. Howard, "The State University in America," 340.

in contempt (and vice versa), agricultural education was not important until land values increased, and farmers wanted practical information, whereas colleges could provide only theory. Scott says land-grant colleges "were painfully aware of the general failure of their schools to influence effectively any large number of ordinary farmers." Many viewed the agricultural colleges with contempt, and by 1900 the colleges were desperate to gain rural support, if only to lure farmers' sons to enroll in the poorly attended programs.[3]

The expansion of land-grant colleges and universities across the West in the 1890s required the development of publicly funded high schools to provide the necessary preparation for higher education. Across the West, the colleges were often put in place by outside forces, while the local school districts had to bring their high schools up to speed to prepare students for college. The first students enrolled at several land-grant schools in the West were actually taking preparation courses in order to begin college classes. It was not a situation where many public school students had completed high school and were clamoring for higher education.

The focus on college-age students was a shift for agricultural colleges, which had been targeting workshops and field days to adult farm operators. Federal, state, and local institutions sorted out whether and who would teach vocational or academic subjects, and to whom. Would adults, children, the poor, or country or city people receive public education? Frederick Jackson Turner espoused the state college as the new frontier, once land became scarce, but critics were not always enamored with the idea of spending tax dollars on education that had no practical value. According to historian Richard Kirkendall, farmers thought that tax-supported efforts to teach agriculture were a waste, that book farming was ridiculous, and that the colleges were providing useless information taught by people who knew nothing about growing plants and animals. They lacked confidence that higher education would prove its worth.[4]

Farmers challenged the colleges, finding their programs, ideology, and delivery techniques wanting. Instead, rural people embarked upon their own educational programs. In the South, the Alliances began

3. Scott, *The Reluctant Farmer: The Rise of Agricultural Extension to 1914*, 29, 138.
4. Kirkendall, "The Agricultural Colleges: Between Tradition and Modernization," 11.

lecturer-based community gatherings where attendees discussed agriculture, particularly related to economics, and finance related to the crop-lien system. Across the Midwest and Pacific Northwest, the Granges, too, sponsored a lecturer system, disseminating information to local groups who gathered to learn the latest findings related to farming and rural life. In New England and the Northeast, the Chautauqua movement promoted the ideal of self-education along with intensive reading and public lectures. The Chautauqua Literary and Scientific Circle began in 1878 in New York State and within a few years had more than sixty thousand people enrolled in correspondence courses, with many others attending seasonal readings and meetings across the country.[5]

Farmers' institutes—gatherings of farmers to hear lectures on agriculture and business topics—had been popular before the Civil War and grew in popularity afterward. Connected with colleges or agricultural societies in the early years, they were eagerly embraced by railroads in the late nineteenth century. In some cases, free institutes were coordinated by the local college and funded by the railroad; in others, they were organized by a state board of agriculture and held as "public meetings." Charging those who attended a small fee usually paid expenses. Topics covered included stock breeding and management, fruits and fruit growing, farm architecture, farm accounts, crop raising, and farm engineering. Eventually, state legislatures appropriated funds to hold some institutes and printed bulletins to augment information presented at the sessions.[6]

The Grange and Alliance lecture system and the agricultural press stimulated farmers to seek information and facts and to promote fairs, exhibits, libraries, and reading circles. Farmers wanted more information and were seeking it in a variety of places. In such a decentralized system, where every man (and woman) could be an expert, where inexpensive books and pamphlets were widely available, and where no one paid for information, how could the fledgling agricultural colleges create a place for themselves?

As historian Alan Marcus has argued, the new land-grant colleges aimed for the same goals farmers had: to create a new sort of farmer.

5. Lawrence Goodwyn, *Democratic Promise: The Populist Moment in America;* Alfred Charles True, *A History of Agricultural Extension Work in the United States, 1785–1923,* 43.

6. True, *History of Extension Work,* 11.

Yet conflict arose over exactly what this new sort of farmer should be. Farmers essentially wanted agricultural business schools, in which their children could learn practical applications of up-to-date management tools. For farmers, this meant a hands-on approach. Colleges, in the farmers' view, were places where successful individuals would pass on their experiences and wisdom to students who would return to their homes ready to take up the reins from the previous generation in agriculture. For both farmers and colleges, agricultural education centered on the college farm, where students could perform predictable tasks while learning new information from men who had already proved themselves experts by succeeding at farming themselves.[7]

Roy Scott points out that the agrarian organizations that rose out of the wave of agrarian discontent in the latter nineteenth century eventually strengthened farmers' support for land-grant colleges. Unwilling to settle for a classical curriculum, the Grange, Alliances, and other groups pushed the colleges to adopt extension work as a way to validate their programs and gain support from the public. Institutes and extension education offered a vehicle to link the colleges with the adult farming community.[8]

The agricultural colleges, however, had quite a different vision. They saw agricultural colleges as places where scientists identified principles that became bodies of knowledge inaccessible to the general public. That knowledge base provided the platform for expertise, which would be dispensed to the public through college-controlled programs. As Alan Marcus notes, "Only scientific principles could improve farming and only scientists could deduce them." Farmers could be taught how to recognize such principles, and how to put them into practice, but they could in no way identify and explain them alone. Scientists would lecture; farmers would absorb the wisdom and put it into practice. To agricultural colleges, farmers were a "dependent caste," one that both needed and respected what scientists could do for them.[9]

As educational programs grew in popularity, colleges and farmers attempted to minimize the other: colleges wanted farmers in a dependent

7. Marcus, "The Ivory Silo: Farmer-Agricultural College Tensions in the 1870s and 1880s"; Kirkendall, "Agricultural Colleges," 6; D. Sven Nordin, *Rich Harvest: A History of the Grange, 1867–1900*.

8. Scott, *Reluctant Farmer*, 37, 38.

9. Marcus, "Ivory Silo," 29.

role; farmers refused to accept the knowledge coming from the colleges, referring to it disparagingly as "book farming." The conflict was formalized through the Morrill Act of 1890, which reinforced the position of both sides. It funded agricultural education while defining and directing how such education would proceed: farmers had oversight of the programs through federal control, and scientists were reinforced by the act's definition of agricultural education as based in the sciences and their application. Although the act provided something for everyone, at the same time it only muddied the waters. The problem was with science itself. Reality had not kept pace with expectations, and while agricultural research accelerated, results, implications, and methods for applying new findings were slow to develop. In some cases, research findings simply reinforced traditional knowledge, leading farmers to view research as worthless at best or to suspect that researchers deliberately withheld valuable information.

Farmers' Institutes

Agricultural education divided into two main camps: those who sought to teach the application of practical science to farmers and those who wanted colleges to educate students to become scientific investigators rather than hands-on farmers. To the latter, foreign languages like French and German were essential, along with learning laboratory procedures and scientific methods. Experimenters would identify scientific solutions, which the masses would apply. Educators identified with becoming practicing scientists, searching for knowledge beyond the confines of the farm. On the other hand, farmers wanted practical educations. For decades, the situation created a distrustful atmosphere surrounding the colleges and how they taught agriculture. To bridge the gap, the agricultural colleges organized farmers' institutes of their own as a way to showcase scientific research to local farm groups outside the college classroom. These institutes were directed not toward enrolled students in the colleges, but to the general public, recruited through mailings and announcements in local newspapers. Beginning in the 1870s, farmers' institutes competed with the Alliances, Granges, and, in some cases, Chautauquas for farmers' attention, but they focused on science's utility and importance to agriculture. They were created to boost agricultural scientists' image to the public, but, bowing to agrarian critics, the institutes evolved into institutes where college experts as well as local farmers spoke. To compete with the growing number of public experts, colleges began offering short courses, bringing farmers onto campuses where the academics controlled the agenda.

The best example of agricultural colleges linking with public education appeared at the University of Wisconsin. Wisconsin established an agricultural experiment station in 1883, which became part of the new state college of agriculture in 1889. Research at the experiment station led to Stephen Babcock's butterfat test for milk (which he did not patent, making the invention public), which alone saved the state's dairy industry eight hundred thousand dollars per year, twice the operating budget of the whole university. The test allowed milk producers to determine the fat content of milk, obtaining payment for milk depending upon its grade. The test alone generated substantial public support for the university, but the school's wide array of outreach programs reinforced its mission to educate a broad audience. A "Wisconsin system" developed, whereby formal institutes were funded by the state legislature and organized through the state university's board of regents. Wisconsin's widely publicized institutes combined school and conference features, meeting at locations where they were invited by farmers' petitions. Sessions consisted of fifteen-minute lectures followed by ample time for questions from the audience. Speakers from industry as well as the state agricultural college presented material. In 1905, the superintendent of the institutes explained that "the conductor at each meeting promptly shuts off all partisan political discussions or statements based on ignorance, prejudice, or superstition. Charts are used extensively in all discussions. Models and animals are also sometimes used." *World's Work* explained the popularity of the Wisconsin program, saying, "There is something inspiring in this thought of a university with the whole state for its campus and the whole population for its student body. It is a university living up to its name." The agricultural extension division brought "the University directly in contact with almost every farmer and farmer's family in the entire state."[10]

Institute sessions lasted two days, and women participated from the beginning. Sessions took on a social as well as an educational flavor,

10. J. Hamilton, "History of Farmers' Institutes in the United States," 18; Frank Parker Stockbridge, "A University That Runs a State," 706, 704.

giving many farm couples a chance to socialize for a few days, similar to the county fair. Even better, the institutes were held when farmers had seasonal breaks in labor. Farmers' institutes conducted by colleges swelled in popularity by the turn of the century, with institutes held in forty-seven states in 1899, holding two thousand institutes with attendance of more than five hundred thousand farmers.[11]

In 1896, the American Association of Farmers' Institute Workers was organized in Wisconsin, and for several years the group met and discussed possible federal funding of the institute programs through land-grant colleges. In 1901, a representative of the Department of Agriculture from the Office of Experiment Stations joined the group. Eventually, with passage of the Smith-Lever bill, the farmers' institutes movement would fade as their work was taken up by the extension programs of agricultural colleges.[12]

Wisconsin also pioneered successful short-course programs, teaching twelve-week-long sessions on a variety of agricultural science topics, as well as a comprehensive correspondence study program in a variety of academic subjects. The university began short courses though the Department of Extension Education in the 1880s; by 1904, dozens of universities sent emissaries to see what Wisconsin was doing and to copy it at home. Indeed, Spillman had availed himself of the University of Wisconsin's short-course program to get up to speed before beginning his teaching assignment at the Washington State Agricultural College and School of Science in Pullman.[13]

Because railroads were instrumental in developing the West, it was only logical that they were at the forefront of agricultural extension, too. Railroads were an integral part of farmers' institutes in many regions of the country. They played a key role in the development of popular agricultural education, beginning with their outright promotion of railroad lands for settlement by farmers and their special trains that visited farming communities to exhibit new seeds, crops, or techniques. By giving free transportation to speakers and annual free passes to land grant–college officials, as well as excursions for farmers, they tried to boost agricultural production along their routes as well. In 1904, two

11. True, *History of Extension Work*, 20.
12. Ibid., 25.
13. Marcus, "Ivory Silo," 35; Kirkendall, "Agricultural Colleges," 10; Cremin, *Transformation of the School*, 167.

railroads in Iowa, working with the agricultural college, ran a "*Corn Special*" that included lecturers, exhibits, books, bulletins, and demonstrations. It stopped at stations where crowds gathered to listen to lectures and pass through the train-car exhibits. Train exhibits flourished across the country, with one day's attendance in Illinois estimated at 3,500. Freight agents acted as agricultural advisers, telling farmers where crops were being grown, yields, and market potential. One line gave out eight hundred purebred bulls and six thousand purebred pigs and offered prizes for the best-managed farms along its line.[14]

There were also "independent institutes" operated by the Granges, Chambers of Commerce, and other organizations. By 1914, there were 1,643 independent institutes in eighteen states, with an attendance of 345,509.[15]

Along with lectures, Chautauquas, and farmers' institutes, a fourth educational venue emerged at the turn of the century to cater to the educational needs of rural people: the Movable School. In 1896, Dr. George Washington Carver was hired to head the Department of Agriculture at the Tuskegee Normal and Industrial Institute. Tuskegee reached out to the community with lectures and weekend trips to small communities where faculty members gave demonstrations. Black farmers also traveled to workshops on the Tuskegee campus, but many were unable to get away from their farms for any sort of educational program, so Carver realized he needed to connect with rural farmers through educational outreach on a steady basis. Along with Booker T. Washington, he organized a stagecoach to carry a group of lecturers and demonstration materials to country dwellers on the weekends. That evolved into a large two-horse traveling wagon, with sides that opened to reveal a small milk separator, a churn and butter-making tools, and charts on soil building, orcharding, stock raising, and other operations. Faculty members traveled along and lectured while showing how to operate the various pieces of equipment. The wagon was funded by a New York banker and philanthropist, Morris K. Jesup, so the wagon was called the "Jesup Agricultural Wagon." Eventually, the wagon and demonstrators took plows and planters, cotton choppers, seeds, samples of fertilizers, milk testers, and other accoutrements of scientific agriculture to

14. Scott, *Reluctant Farmer,* 171; True, *History of Extension Work,* 29.
15. True, *History of Extension Work,* 35.

the countryside. The "Movable School" made the rounds in a community, then settled down for a session of outdoor lectures and demonstrations before a gathering of farmers.

In 1906, the Movable School contacted two thousand people per month, blacks and whites, on small farms or tenants on large ones. In 1906, the General Education Board, a Rockefeller family philanthropy, began funding the Tuskegee program. The first "cooperative extension" program in the United States emerged at the Tuskegee Normal and Industrial Institute, and T. M. Campbell, a black agent, became the first "cooperative extension agent." (Coincidentally, a white agent, W. C. Stallings, was hired the same day as Campbell, to work as an agent in Smith County, Texas, but Stallings served only one county.) Although the program had much promise, funding did not expand the program. In 1914, the wagon wore out and there were no funds to replace it, so the school began using the train to send out demonstrations. In 1918, the school received funds from the state agricultural extension office to purchase a truck, named the "Knapp Agricultural Truck." It wore out in 1923, so local black farmers contributed to purchase a new truck, named the "Booker T. Washington Agricultural School on Wheels."[16]

Science and Education

Although Pasteur's work had realigned the public's thinking about the value of scientific work, in the 1890s little else of real practical value had come from scientific agriculture. The public was not far off with criticism of agricultural science; the disastrous "rain follows the plow" theory—that once land was tilled, dry climates would become more humid—was still fresh in the public's mind. That had cost a generation of families heartbreak and loss trying to turn the great American desert into well-tended farms. The USDA itself had been a laughingstock as animal scientists scrambled on the ground, looking for "sporules" they believed caused Texas tick fever in cattle, something that cattlemen had long connected with ticks but could not be explained until Howard Ricketts's pioneering work on tick-borne fever in Montana was applied to typhus. Scientific solutions were complex, and although

16. B. D. Mayberry, "The Tuskegee Movable School: A Unique Contribution to National and International Agriculture and Rural Development," 85–94.

colleges and scientists lauded the potential, before 1900 agricultural science held far more potential than actual solid knowledge.[17]

Therefore, it comes as no surprise that the agricultural experiment stations were established on rather weak scientific ground. David Danbom and Margaret Rossiter note that the stations were established before there was adequate science on which to base them. Chemistry and botany were the only solid fields, and they were undergoing change. Without much to base their expertise upon, the experiment stations and scientists engaged in research appeared to represent political patronage. Danbom notes that professionalization influenced the agricultural experiment stations, where public pressure from client groups threatened to undermine their authority.[18]

As science held great promise for the future, so too did education. In many cases, education became a panacea for rural discontent. Just as urban dwellers identified with education as an avenue upward in an increasingly industrial world, rural residents looked to higher education for opportunity as well. Frederick Jackson Turner pointed to state-funded higher education as a new frontier in his essay "Pioneer Ideals and the State University," delivered at Indiana University in 1910 and later published in a collection of essays, *The Frontier in American History*, in 1920. At a time when farm prices were high and land prices were escalating, farm youth found it difficult to gain access to land they could farm. The "farm ladder," the idea that one began working as a hired laborer and eventually moved into landownership, no longer signified a path to success (defined as farm ownership) for young, or even older, farmers. With no more "free land" available to the industrious, state universities offered an alternative to the agricultural ladder for bright, ambitious rural youth. With knowledge, perhaps they could overcome an increasingly restricted system.[19]

---

17. Richard White, *It's Your Misfortune and None of My Own*, 132; Cecil Kirk Hutson, "Texas Fever in Kansas, 1866–1930," 77; Claire Strom, "Texas Fever and the Dispossession of the Southern Yeoman Farmer"; Joseph G. McCoy, *Historic Sketches of the Cattle Trade of the West and Southwest*; Victoria Harden, *Rocky Mountain Spotted Fever: History of a Twentieth-Century Disease*.

18. See Danbom, "The Agricultural Experiment Station and Professionalization: Scientists' Goals for Agriculture"; and Rossiter, "The Organization of the Agricultural Sciences."

19. Turner, "Pioneer Ideals," in *Frontier in American History*, 280–89; Grant McConnell, *The Decline of Agrarian Democracy*.

By embracing science and extension education, agricultural colleges gained public support. For farm parents, schools, especially agricultural colleges, promised to provide their children with tools—science, in particular—that would allow them to remain on the farm and in the rural community. The extension model for agricultural education, which taught adults in their own community, emerged not from colleges but from public educational programs. Farmers' institutes, emerging from the Farmers' Alliances and Grange lecture programs, grew in popularity until by 1890 twenty-six states held farmers' institutes or similar public meetings to disseminate information on a permanent basis. By that time, most were connected with state colleges and universities and faculty served as lecturers. The sessions usually lasted two days and were held in winter when farmwork was light. In 1896, the farmers' institute movement led to the formation of the American Association of Farmers' Institute Workers in Wisconsin. In 1897, the group discussed the idea of getting federal aid to land-grant colleges for agricultural extension work. That same year, President William McKinley tabbed James Wilson, a multitalented college administrator, former Iowa farmer, and former Republican congressman, to serve as secretary of agriculture. Wilson went on to serve under three presidents, as he was in Roosevelt's and Taft's cabinets as well. In 1901, Wilson requested five thousand dollars for the Office of Experiment Stations to promote farmers' institutes, and the federal government entered the institute business.[20]

Spillman as Educator

The challenge to agricultural education came in two forms: how to reach farmers and what to tell them. The issues were content and delivery, and, as a scientist and educator, William Spillman wrestled with both. Before his move to the Department of Agriculture, his career as an educator encompassed a small Missouri schoolhouse, college classrooms, and farmers' institutes. He also taught in what might be termed nontraditional venues, too, as a lecturer at Grange meetings and a frequent speaker at Chautauqua gatherings. He had experience teaching

20. John A. Garraty and Mark C. Carnes, eds., *American National Biography*, s.v. "Wilson, James"; True, *History of Extension Work*, 27.

in formal educational programs, as well as in independent popular ones. Although his intellect and enthusiasm for education, agriculture, and science undoubtedly disposed him to be an educator, he came of age during an era when education was becoming central to culture and society. New theories and educational models were tried, discarded or adopted, and in some cases strongly contested as Americans sought to both prepare for the modern era as well as minimize the impact of industrialism. Spillman typified this era, as he continually strived to identify important problems for farmers, formulate solutions, and then disseminate the information so farmers could make better decisions. Like many Progressives, he found his theories and ideas sometimes needed revision; if so, he was undaunted, pushing off in a new direction, continually seeking efficient resolutions to problems he encountered.

Integral to creating curriculum and content for any agricultural education program was the need to identify what was essential for success in farming. Access to capital was important, but education had to expand beyond that. Defining what it meant to be a successful farmer and identifying a systematic process for becoming one were central to developing curriculum and content in agriculture education. Spillman researched success and the steps it took to get there, applying scientific methods as he shaped a philosophy of farm management during an era of continual change. He defined what he considered to be the secret to successful farming, which could be applied as a philosophy of life going far beyond agriculture. He believed that even novices could be successful farmers, despite all the hurdles, if they understood how to make the right decisions. Assessing the approaches and practices of those farmers who were unusually successful, he surmised that "perhaps the most important element in the man is the ability to distinguish between a good suggestion and a poor one."[21]

How to do that assessment was the question that drove Spillman's work as he set out across the country to gather information about farming, systematically assess its value, then disseminate the results to farmers so they could improve their practices. His efforts reflected the new interest in economic questions related to agriculture that had developed in the educational institutions. He was an agrostologist (agronomist),

21. Spillman, "General Farming" (1904), quoted in R. Spillman, "Biography of Spillman," MASC, 217.

but his work centered on conducting and supervising surveys and studies. Rather than creating a set of top-down directives for farmers to adhere to, he saw individual free choice, based on solid data and professional interpretation, as the cornerstone for agricultural success. How to put that philosophy into practice was unclear, but he thought it would take the scientific method—experimenting, assessing, refining, and rethinking data—in order to provide a firm basis to farmers for their decision making.[22]

Spillman's thinking was influenced by genetic science as it stood in the early twentieth century, heavily rooted in Darwinism and natural selection. The thinking that superior samples needed only to be identified, then replicated, was rooted in Mendelian ideas. In the *Journal of Farm Economics,* C. B. Smith remarked on Spillman's legacy by noting his innovative idea about identifying superior farmers, studying them, and promoting their practices. Smith described Spillman's philosophy: "In any group of one hundred farmers, taken as they come anywhere in the United States, there are between fifteen and twenty farmers whose average labor income is approximately four times the average labor income of the whole group. This law is as significant in extension work as Mendel's law is in plant breeding." This sort of thinking, he added, "points the way to readjustments that may be made by farmers in each community, anywhere, looking toward a betterment of their condition." One might compare Spillman's ideas about identifying successful individuals and concentrating efforts to improve agriculture on them with a contemporary thinker, W. E. B. Du Bois, who held the same ideas, proposing that the black race could be uplifted by focusing on the "talented tenth." Perhaps Spillman was searching for the talented tenth among farmers; certainly, he and Du Bois shared a common agenda, melding science and sociology, hoping to uplift a group by focusing on the ablest cohort.[23]

Fundamental to Spillman's work were two ideas he developed. One was the idea that "every farm is an experiment station," where farmers are experimenting on their own to work out solutions to problems in a

22. Gladys L. Baker et al., *Century of Service: The First 100 Years of the United States Department of Agriculture,* 44.
23. R. Spillman, "Biography of Spillman," MASC, 272.

practical, businesslike manner. Second, analyzing the practices of farmers who succeeded could reveal important facts valuable to the rest of agriculture. His approach suggested answers to broad questions unlikely to be resolved in a test plot and garnered immense public approval because the farming public felt incorporated into progress, rather than sidelined. He saw farming as a series of interdependent decisions to be made by the farmer based on tradition as well as science. Land was a resource, and the end of "free land" meant farmers had to preserve its fertility for the future. His goal was to help figure out how to best manage their increasingly scarce land. Spillman's solution was uniquely his: study what successful farmers in a particular region do and replicate it. In an era of "experts," his principle was almost heresy. Yet it succeeded and provided both a way for agricultural researchers to garner knowledge directly from the field as well as a way to show respect and recognition for farmers, who never forgot that Spillman was both a scientist and their friend.[24]

In 1902, Spillman began developing cooperative agriculture projects with fifteen experiment stations, as well as five model demonstration farms that he coordinated with Seaman Knapp. Knapp was a significant figure in southern agriculture due to his widespread program of demonstration farms (which will be treated more fully below). While Knapp moved from model farms to his southern demonstration work, Spillman continued to work with the model farms. Spillman established twenty-two in 1904 and twelve more in 1905, in the states of South Carolina, Georgia, Alabama, Mississippi, North Carolina, Arkansas, Louisiana, and Texas (which had the most, twelve). As the Bureau of Plant Industry's agrostologist, he worked with the various state experiment stations with the goal of increasing soil fertility while shifting the farmer away from monocultural cotton growing. Holding farmers' institutes right on the farms was effective in obtaining public interest; the meeting at the Uniontown, Alabama, project resulted in a thousand farmers in attendance. The farms exhibited techniques and crops such as alfalfa, sorghum, corn, peas, and the inevitable cotton. Spillman focused a lot

---

24. E. H. Thomson, "The Origin and Development of the Office of Farm Management in the United States Department of Agriculture," 14–15; R. Spillman, "Biography of Spillman," MASC, 198.

of attention on livestock, with hogs, cattle, and goats in the spotlight. One farmer who participated in the model-farm effort claimed it was the most profitable farming he had ever done because of the diversified approach.[25]

Spillman's model farms differed from Knapp's work in that Spillman promoted diversified agriculture, whereas Knapp concentrated on getting higher cotton yields. Spillman hoped to break the cotton monoculture in the South by revolutionizing southern agriculture. In spite of the apparent success of the model farms and institute program, southern farmers did not adopt diversification. Spillman admitted the program had failed. Southern conditions made wheat and hay difficult; the humidity and lack of transportation to a market made those crops prohibitive. Livestock—always the backbone of Spillman's agricultural philosophy—were susceptible to internal parasites as well as the problematic cattle tick so prevalent in the southern climate.[26]

By the end of 1904, Spillman believed that the huge amount of responsibility (and work) he held in the Bureau of Plant Industry necessitated creation of a separate bureau within the department, and he urged creation of a division under his own management, with appropriate prestige and funding. An offer from the Pennsylvania State College to accept the position of dean of the School of Agriculture arrived at the same time. He turned down the position, ostensibly because Mattie had no inclination to adopt a lifestyle that included social obligations connected with the role of a college dean's wife. Perhaps the secretary learned of Spillman's dissatisfaction or the Pennsylvania job offer; in any case, the response was quick. Spillman was given the position of chief of the new Office of Farm Management (OFM), officially launched on July 1, 1905. It was not exactly a bureau, as Spillman had hoped, but it was an independent entity within the Bureau of Plant Industry. From its inception until 1910, the Office of Farm Management focused on farm practices, publishing information about forage crops, dairying, range-pasture reseeding, model farming, and restoring depleted soils.[27]

25. Scott, *Reluctant Farmer*, 257; R. Spillman, "Biography of Spillman," MASC, 221–22.
26. Scott, *Reluctant Farmer*, 259; R. Spillman, "Biography of Spillman," MASC, 266.
27. R. Spillman, "Biography of Spillman," MASC, 228.

The Office of Farm Management

Secretary James Wilson stated that he created the Office of Farm Management as a reorganization of the Bureau of Plant Industry, which would bring together all the facts developed in the bureau as a whole, sifting the results and applying them in a practical way where they would do the most good. To inaugurate research, a survey of successful—"best-paying"—farms ensued, assessing management practices, market access, landownership or tenancy, soils, and climate. The scope was enormous, something that Spillman must have adored, as his interests and talents were so varied.

The Office of Farm Management was organized to study regional farming practices and management problems and to serve as an arm of the department to work with colleges of agriculture and farmer organizations. The office would change names and focus several times until 1922, when it became part of the Bureau of Agricultural Economics. William Spillman was the right man to run the office, with his background in college teaching and farmers' institutes, as well as recent research and publications on farm management, model farms, and opportunities in agriculture for the general public. The department under Wilson in those formative years was notable for its rather decentralized administration. Wilson believed in hiring extraordinary people, then letting them identify problems and figure out how to solve them. For the broad scope given the Office of Farm Management, Spillman was well suited.[28]

Spillman loved gathering data, interpreting them, and making sense of them, but he also enjoyed presenting them to the public, and the 1904 St. Louis World's Fair gave him an opportunity to take his message to a large number of people. To represent the department, he designed and developed a huge map of the United States, covering five acres on a sloping hillside outside the Palace of Agriculture. Each state on the map was planted in its major crops, in areas proportional to their share of the state's acreage. The map could be viewed up close, where one was able to examine the plant specimens firsthand, as well as from above, particularly from high on the Ferris wheel, which made

28. See Thomson, "Origin and Development."

its debut at the fair, too. As his son, Ramsay, later remarked, "My father had a knack for the graphic presentation of statistics."[29]

That knack also led Spillman to develop a unique visualization of statistical data: the dot map. Dot maps would eventually become a mainstay of departmental statistics and be accepted in other fields as well. It was a simple way to show geographic distribution of statistical occurrences: a dot on the map represented a specific amount or population. The number of dots clustered represented distribution of a commodity or population. Ramsay remembered his father bringing work home to their modest flat in Washington, D.C. (they lived on the top floor of a house at the time), where he assisted his father in marking maps for the department. Spillman brought home a map of counties on which he stamped dots using the sharpened end of a wooden penholder, inked on a stamp pad. Eleven-year-old Ramsay read aloud statistics from the 1900 census report while his father stamped maps to signify those amounts. For example, a county that grew thirty thousand acres of corn would be stamped with three dots, one for each ten thousand acres. The first publication of an agricultural dot map was in Spillman's book *Farm Grasses of the United States*, published by Orange Judd in 1905.[30]

Spillman was thrilled to have his own office to manage, and he operated it much like a college classroom. Most of the investigators working for him were young male college graduates who called him "Prof" and frequently dropped in at the Spillman home to talk about the work as well as personal matters. He guided their work skillfully, pushing them to take on challenges, allowing them to take credit when it was due, and encouraging them to succeed with praise and recognition.

Billy Sando, a messenger hired at the department in 1909, is an example of Spillman's mentoring. Sando was ambitious, learning what he could on the job. When Spillman's longtime secretary, Nellie Price, was away from her desk, Billy often sat there, performing small tasks to help her out. One skill he developed was the ability to sign documents with a replica of Spillman's signature as well as his initials, something Spillman found very valuable, as he was away from the office frequently. He could discuss letters and documents over the telephone

<hr>

29. R. Spillman, "Biography of Spillman," MASC, 207.
30. Ibid., 207.

with Nellie, then have Billy sign or initial Nellie's typed responses. When Billy began to show serious interest in working with plants and plant breeding, Spillman encouraged him to go to college, offering to lend him the money he needed. Sando's family was reluctant, so Spillman helped craft a flexible work schedule so Sando could keep his job at the office while attending college classes. Eventually, Sando earned a Ph.D. and became an agronomist at the Department of Agriculture's Arlington farm. In 1930, Sando and Spillman coauthored "Mendelian Factors in the Cowpea (*Vigna* species)," a paper published by the Michigan Academy of Science.[31]

Spillman exhibited skill at moving in both the scientific world as well as the practical. In one example, he interviewed J. S. Cates for a position at the Office of Farm Management, asking the fresh young Cornell graduate if he knew anything about botany. Cates squirmed and promised to brush up on the subject. Spillman's response was "Don't. I want a man to work on weeds. Most of the work so far on weeds has been by systematic botanists, and they mount up herbarium specimens of them, and label them, and that's as far as it goes." He told Cates, "I want you to forget about systematic botany, and learn how to kill weeds so you can show the farmers how to do it."[32]

A practical, pragmatic man, Spillman was well liked by farmers and also recognized as a highly successful professional scientist. He was comfortable and interested in working with audiences that ranged from small-town newspapers to Britain's Royal Society. He spent time and attention on both simple farmers' questions as well as arguing major scientific breakthroughs before international audiences. In 1910, two examples reveal the wide range of interests and audiences he addressed. In Spokane, Washington, the local newspaper ran a contest asking subscribers to submit entries for the "Good Farming and Attractive Country Homes" contest; Spillman agreed to judge the entries and spent quite a bit of time perusing the entries. That same year, Temple University in Philadelphia gave him an honorary doctorate of science in recognition of his genetic work with wheat breeding, an honor he was very proud of, later donning the golden yellow hood at convocations during the eleven years he was on the faculty teaching international

31. Ibid., 228, 283.
32. Ibid., 287.

subjects at Georgetown University. Spillman respected and valued his audience, without regard to status or position; his intent was always to reach the audience rather than to impress them.[33]

Although Spillman's diversified model farms in the South failed to motivate farmers to switch from growing cotton, he searched for other educational models that might succeed. The early combination of farmers' institutes and railroad-sponsored tours had been a perennial hit. Railroads were instrumental in organizing and promoting train demonstration tours with traveling exhibits and lecturers who demonstrated to one rural audience after another.[34]

So it was not unusual that George Cullen, a manager with the Delaware, Lackawanna, and Western Railroad, decided to set up a demonstration farm along its line in Broome County, New York. Cullen enlisted the New York State Agricultural College to manage it and went to Washington, D.C., where Secretary Wilson introduced him to W. J. Spillman. Spillman, fresh from his model-farms project, discouraged Cullen from setting up a demonstration farm, suggesting that a county agent be hired instead. Cullen hired an agricultural college graduate, housing his office in the local Chamber of Commerce. The agent was to identify successful local farmers and create programs based around their practices, demonstrating on farmers' own farms. The agent was supported by the cooperation among the railroad, Office of Farm Management, local Chamber of Commerce, Grange, and state college. It was truly "cooperative," in the sense that various groups came together with the single goal of improving local farming.

Out of the Broome County experience, Spillman realized that farm-improvement work needed to be "local, concentrated, and continuous." It also had to avoid farmers' perceptions that something was being done "for them by business and railroad interests, which made them wary the effort was focused on increasing production, while their goal was to get more money for what they produced." According to Alfred Charles True (who later directed the States Relation Service), the first county agents in New York State succeeded because they were also active Grange

33. Ibid., 263.
34. See John Hillison, "Agricultural Education and Cooperative Extension: The Early Agreements."

members. The New York–style county-agent model quickly spread throughout other northern states well before the 1914 Smith-Lever Act created the Agricultural Extension Service. North Dakota counties taxed to support twenty-one county agents by 1914; nine Wisconsin county agents worked with the Office of Farm Management, receiving salaries from the Department of Agriculture by 1914. Well-organized county-agent systems, linking local control with state colleges and the Office of Farm Management, spread to other states before 1914, including Missouri, Illinois, Michigan, Kansas, West Virginia, Idaho, Minnesota, Colorado, Indiana, Washington, Nebraska, Ohio, Massachusetts, Wyoming, California, and Utah.[35]

Whereas Spillman's model for educating farmers was based on the belief that there were superior farmers who already knew what was best and that an expert had only to play the role of intermediary, another farm educator was working to expand a quite different farm demonstration program. Seaman Knapp established a system of demonstration farms and agricultural demonstration agents in several southern states, based on agents going onto farms and showing farmers how to do particular techniques Knapp had identified as successful. His was a top-down model, with the government expert arriving in the county to show the local farmers how to increase their production. Knapp's work is lauded in departmental literature as a commendable union of corporate interests, banks, and farmers, but there is scant evidence of its effect upon farms across the South. Although he called it "diversified" farming, in truth Knapp's demonstration farms pushed farmers more securely into monocrop cotton cultivation.[36]

Knapp had done agricultural education a huge service by instigating the experiment station bill that C. C. Carpenter of Iowa introduced to Congress in 1882 to provide federal funds for experiment stations connected with agricultural colleges. Eventually, the measure passed, in the form of the Hatch Act in 1887. Knapp's reason for connecting the experiment stations with the colleges was that he thought the research would benefit the students, "as object lessons and would perfect and give practical value to the work of the colleges."[37]

35. True, *History of Extension Work,* 78, 79, 87.
36. See Knapp, "Demonstration Work in Southern Farms."
37. True, *History of Extension Work,* 59.

In 1886, Knapp had embarked on industrialized rice growing in western Louisiana for an English land syndicate that was developing and marketing 1.5 million acres of land. Using highly mechanized production methods, he turned Louisiana into the leading rice-producing state. Advertising, railroad tours, and booming land values attracted potential buyers. Knapp became an enormously wealthy developer in Louisiana, creating a large agricultural syndicate that farmed 1 million acres. With the assistance of the Bureau of Plant Industry, Knapp set up several demonstration farms, showing farmers how to grow rice in the region, but local farmers ignored the effort, perhaps due to the huge capital necessary to embark on rice culture, according to historian Deborah Fitzgerald. Knapp switched to marketing the farms to midwestern farmers, using model farms as promotions. Thousands of farmers relocated, and rice growing extended into Texas as well. In 1898, his longtime friendship with James Wilson garnered him trips to Japan, the Philippines, and Hawaii to bring back superior rice seed. In 1901, he went to Hawaii and Puerto Rico as an agent of the department.[38]

In 1903, weevils attacked the rice plantations, so Knapp switched to cotton farming. The cotton boll weevil was on the horizon, however, as parts of Texas had been infested for about a decade, and in time the insect would spread to Knapp's farms. The boll weevil was not to be ignored, particularly at a time when the department was cutting its teeth on entomological research due to the cattle tick fever in the same region. Secretary Wilson obtained $250,000 from Congress to fight the weevil, giving Knapp $40,000 to set up an office in Houston as an agricultural consultant. Through his Houston office, Knapp garnered business contributions to fund demonstration farms and met with "farmers, bankers, merchants, railroad presidents, and other business men." Farmers were reluctant to participate, but Knapp changed that by enlisting local bankers and merchants, who denied credit to farmers who did not "cooperate."[39]

In 1906, David Houston, president of the Texas State College of Agriculture, introduced Knapp to Dr. Wallace Buttrick, director of the

---

38. C. Vann Woodward, *Origins of the New South, 1877–1913*, 120, 410–12; Fitzgerald, *Every Farm a Factory: The Industrial Ideal in American Agriculture*, 14; Judith Sealander, *Private Wealth and Public Life: Foundation Philanthropy and the Reshaping of American Social Policy from the Progressive Era to the New Deal*; True, *History of Extension Work*, 59–61.

39. True, *History of Extension Work*, 60; McConnell, *Decline of Agrarian Democracy*, 29.

General Education Board, a philanthropic arm of the Rockefeller Foundation. Houston was on the board of directors of another Rockefeller philanthropy, the Southern Education Board (SEB). Knapp, then seventy-three years old, was enlisted to direct a demonstration program for the GEB, which would be administered through the Department of Agriculture.[40]

The General Education Board

Progressive educational programs linked with private philanthropy during the early twentieth century, as corporate philanthropists attempted to leave a legacy as well as to effect social change by giving generously to fund schools and public programs. Andrew Carnegie and John D. Rockefeller, as well as others, funded a variety of educational programs. The nation had never developed a centralized educational bureaucracy, something Noah Webster had lobbied for in the eighteenth century, but the idea never caught on in a nation that embraced individualism and local democracy. Without a centralized system, it was difficult to accomplish educational change on a wide scale. The only federal support for broad educational programs was within the land grant–college program, which was minimal until the mid-1910s. The influence of powerful foundations with vast resources and targeted goals for social change made for important players as the nation's educational system expanded.[41]

In 1902, Ida Tarbell's scathing revelations of Standard Oil's business transgressions appeared in a series of articles in *McClure's*, running over a two-year period. She set readers' teeth grinding in anger over Rockefeller's ruthless methods at Standard Oil to monopolize the nation's petroleum industry. Charged with breaking the Sherman Anti-Trust Act, Standard Oil appealed to the Supreme Court and was eventually found guilty and broken into thirty-eight companies. In response, the Rockefeller family embarked on philanthropy, giving funds to secular

40. True, *History of Extension Work*, 59–60; Houston, *Eight Years with Wilson's Cabinet, 1913 to 1920, with a Personal Estimate of the President*, 203; Sealander, *Private Wealth and Public Life*, 47.
41. Harlow Giles Unger, *Noah Webster: The Life and Times of an American Patriot*; Thomas Neville Bonner, *Iconoclast: Abraham Flexner and a Life in Learning*, xvi.

interests for the first time. Their giving had been limited to Baptist endeavors, but they began giving to educational endeavors as well.[42]

The General Education Board, at first called the Negro Education Board, began in 1902 with a million-dollar donation from John D. Rockefeller Sr., the money to be distributed over ten years. Rockefeller Sr. and his son, John D. Jr., met regularly to administer the fund, with the two retaining control of two-thirds of the monies given out. Initially, efforts centered on black education in the South, but the GEB in practice sought to appeal to southern whites. In spite of much publicity for the efforts, the GEB actually had little effect on black education. W. E. B. Du Bois heartily disagreed with the GEB's policies, criticizing them for segregating schools, focusing on manual training and not education, and catering to southern whites. Ninety percent of the money earmarked for black education efforts actually went to white schools or for medical education. Between 1902 and 1919, the GEB's total funding to black colleges was twenty-five thousand dollars to the Hampton Institute. In what was essentially philanthropic paternalism, northern interests moved to redeem the laggard South by centering on propaganda and publicity for the cause of southern education. But after vigorous and enthusiastic efforts, southern education did not improve much. On the other hand, cotton farming was prospering in the decade after 1900 with record high prices for cotton. High commodity prices pushed up land values, escalating southern farmland values by 103 percent. The cause of southern agriculture beckoned with greater promise than southern education. The GEB shifted its identity to the farm, choosing to focus not on teaching young people in public schools, but on contacting practicing farmers. Agricultural demonstration work aimed at enhancing the productivity of southern agriculture emerged as the next philanthropic enterprise that the Rockefellers pursued.[43]

Literature produced by the GEB later extolled the program as a successful model for foundation philanthropy: foundations fund projects that are proven successful enough to move on to becoming public policy, supported with public funding. In 1915, the GEB was a success,

---

42. Ron Chernow, *Titan: The Life of John D. Rockefeller, Sr.*, 483.
43. Ibid., 486; David Levering Lewis, *W. E. B. Du Bois: The Fight for Equality and the American Century, 1919–1963*, 28. Between 1915 and 1930, almost $21 million was given by the GEB to black colleges; on the bulk of the funds given between 1927 and 1930, see ibid., 286; Woodward, *Origins of the New South*, 402, 407.

a "perfect illustration of the valuable part that can be played by private beneficence," according to director Abraham Flexner. A later history of the GEB, written by Raymond Fosdick and published by the Rockefeller Foundation, concurred. As historian Andrew Morris notes, these men viewed the GEB's work as pioneering efforts in establishing successful philanthropic programs. Their writing, however, "downplays the single-mindedness of the GEB's participation and it obscures the conflict which eventually ended it," according to Morris. The GEB-USDA collaboration was, in Morris's words, "the center of a heuristic and bureaucratic battle that was supercharged with the swirling anti-Rockefeller sentiment of the 1910s."[44]

Knapp's Demonstration Farming

The boll weevil threatened to ruin southern cotton growing and was spreading across the South. Threat of infestation and resulting poor crops created fear in local merchants who refused to extend credit to tenant farmers. Unable to borrow against a future crop, tenants could not plant and began leaving the land. Landlords, tenants, and merchants all sought a quick remedy for the insect that threatened to depopulate the South. The resulting crisis provided an opportunity for the Department of Agriculture to prove its worth to farmers. If the boll weevil could be stopped by USDA expertise, federally administered scientific agriculture could gain public support. When the department learned of Knapp's methods, it had a promising mechanical—if not scientific— solution. Seaman Knapp demonstrated how the weevil could be eliminated through farming techniques. He advocated intensive farming: by plowing and cultivating fields early and often, the life cycle of the weevil in the soil could be interrupted. Intensive farming meant more trips back and forth over the soil with the plow and cultivator—and the adoption of tractor farming by those who could afford it. Knapp worked seeming miracles on lands planted in cotton the past thirty years when he applied heavy amounts of fertilizer. Results were dramatic; even with boll weevil infestations, the crops yielded significantly more than

44. *The General Education Board: An Account of Its Activities, 1902–1914,* 60; Andrew Morris, "The General Education Board and the U.S.D.A.," 16.

before Knapp's demonstration. Knapp's methods did nothing original; fertilizer alone boosted production with visible results.[45]

James Wilson wanted to extend the USDA's farm demonstration program across the South by using Knapp's techniques, and the GEB promised to provide monthly checks to cover expenses. Beginning in 1906, the GEB eventually funneled millions of dollars directly to the Department of Agriculture. The strategy worked. By 1912, one hundred thousand farms across the South practiced intensive cultivation of cotton, applying commercial fertilizers and tilling the soil six or more times per season, under the direction of seven hundred agents.[46]

Meanwhile, the Office of Farm Management under Spillman's direction extolled the virtues of crop diversification and soil conservation; the GEB efforts across the South promoted neither. Southern farmers, and particularly black farmers, wrestled with increased tenant farming, inability to obtain loans and credit, and lack of marketing standards, which allowed buyers to downgrade their products, paying less than farmers deserved for their cotton. The GEB, however, believed that what the farmers most needed was increased production, mechanization, and a better road system. According to the GEB, both black and white farmers profited by rising land values (definitely *not* an asset to tenant farmers). The GEB focused on "the increase in farm productivity by the introduction of machinery and better methods of farming; the general introduction of conveniences and amenities through the telephone, good roads, rapid transit, free delivery, and the parcel post." Transportation—as connected to agriculture—was the GEB's focus.[47]

In the Department of Agriculture, the GEB influence was heavily entrenched. From inside the department, William Spillman began to question the alliance between industry and government, criticizing the GEB's influence. Called the "fairest and squarest man I ever knew and entirely too brilliant a man for Gov't Service" by a colleague at the time, Spillman challenged the working relationship between the Rockefellers and the federal government. Spillman, as head of the Office of Farm Management, was charged with investigating methods of farming and conducting demonstration work that taught improved meth-

---

45. Scott, *Reluctant Farmer,* 212, 211, 217.
46. Morris, "General Education Board," 16; Chernow, *Titan,* 487; *The General Education Board: Account of Activities,* 61; Scott, *Reluctant Farmer,* 226.
47. *General Education Board: Account of Activities,* 194.

ods and systems of farm management. The focus was on discerning appropriate scientific methods for different regions of the country and bringing that information to farmers. The emphasis was on crop rotation, tillage practices, and soil amendments. Spillman, directing a "splendid group of keen young men," according to one participant, inspired hard work and loyalty in those who served under him. Such cohesiveness was necessary because the Department of Agriculture was split between warring factions who battled over departmental turf and agricultural orthodoxy as well as the GEB money.[48]

Congressional Intervention

In 1913, David Houston, a member of the board of directors of the Southern Education Board, a subsidiary of the GEB, left Texas to replace James Wilson as secretary of agriculture. Soon, relations between Houston and Spillman became strained, particularly over the GEB role within the department. "By this time conditions in the department had become so unbearable that I decided to take a hand in helping to remedy them," Spillman later explained. "Accordingly, I wrote a resolution removing the Rockefeller funds from the department amounting at that time to $660,000 a year, and substituting federal funds for them. The resolution also prohibited the Department from co-operating with the General Education Board, or any similar organization." C. Vann Woodward notes that the GEB and Southern Education Board, as well as the Rockefeller Foundation, the Jaines Foundation, and other southern philanthropies during these years, had interlocking directorates. Several members of the boards of directors served the same organizations. As he points out, philanthropy was in the hands of a few individuals, although it was funneled through a variety of organizations.[49]

Spillman enlisted a valuable ally in his fight to eliminate the GEB from the department, Republican Iowa senator William S. Kenyon, who introduced the whole matter to Congress the following year as part

48. Harry B. McClure to Ramsay Spillman, January 10, 1933, box 6, file 53, Spillman Papers, MASC.

49. McCann, "Secretary Houston Attacked as Rockefeller Subordinate," *New York Globe and Commercial Advertiser,* March 28, 1919; Woodward, *Origins of the New South,* 402.

of the agricultural appropriations bill. Remembering what became "a very spirited debate," Kenyon later noted that, indeed, Spillman "played a great part in that beneficent legislation."[50]

That debate opened on April 1, 1914, when Senator Kenyon requested an examination of the General Education Board's role at the Department of Agriculture. Secretary Houston provided Congress with a copy of the memorandum of understanding signed in 1906 by the Rockefellers and then secretary of agriculture James Wilson, which explicitly gave the GEB control of demonstration work in southern states that were not already part of the federal boll weevil–eradication program. The GEB agreed to pay direct and indirect expenses for farm demonstration work, including salaries, and the USDA agreed to pay $1 per year salary to individuals working for the GEB program so they could maintain franking privileges through the department. The GEB's territory for farm demonstration work included North Carolina, South Carolina, northern Georgia, Virginia, West Virginia, and Kentucky. Other southern states remained under the funding and control of the USDA. Houston revealed that the GEB financed positions for twelve economics professors at colleges and universities as well as clerical workers, farm demonstrators, and organizers of youth clubs. The issue expanded, however, when records of the GEB were presented to Congress. Pages of documents revealed that the GEB had six hundred people on the USDA payroll, at $1 salary per year, the rest paid by the GEB. Even Bradford Knapp, Seaman Knapp's son, received the $1 salary along with $625 per month from the Rockefeller Fund.[51]

In an era of impassioned trustbusters, Congress was as wary as the public concerning anything to do with Standard Oil. Senators from Oregon, Iowa, Washington, Kansas, Oklahoma, and North Dakota argued against continuing to allow Rockefeller money to influence government programs. They pointed out that any philanthropist could donate money as he or she saw fit to any charitable cause, and they did not object to the Rockefellers funding education or any other beneficent activity. They just did not think that government agencies should funnel

50. Spillman statement to National Board of Farm Organizations, February 1919, quoted in R. Spillman, "Biography of Spillman," MASC, 311, 312.
51. *Congressional Record,* 63d Cong., 2d sess. (May 23, 1914): S 9152.

such money. Senators from Virginia, Mississippi, and South Carolina defended the GEB involvement, arguing that the money had done much good.

Supporters of the GEB noted that Seaman Knapp had begun demonstration work in the South in a desperate attempt to fight the cotton boll weevil and that, once he gained GEB funding, charitable contributions of $1 million a year were going to the USDA as well as directly funding jobs in southern states. Representative John Small of North Carolina pointed out in the House of Representatives that the arrangement was not a secret. He noted that President Theodore Roosevelt, two secretaries of agriculture, the officials within the Department of Agriculture, and southern people all knew about and approved the funding. Small claimed that he "had it from the highest authority that at all times this cooperation by the board had no strings attached." He said that in North Carolina, the demonstration work had never been political and that the GEB had not tried to influence the work.[52]

Republican Ohio representative Simeon Fess countered Small's praise by pointing out that opposition was based on larger issues—such as the fact that Secretary Houston was a member of the GEB. Small admitted that such was true but that the situation had created a "renaissance in agricultural methods" in the South. Production was up, and farmers were doing better than ever. He focused attention on Knapp's successes, framing the issue around him and avoiding mention of the Rockefeller name. The people in *his* state were not about to criticize the source of such largesse; he stated to applause, "We are neither ingrates nor cowards."[53]

Congressman James Byrnes, a Democrat from South Carolina, pointed out that he supported changes because it was now clear that more than six hundred federal employees had been salaried by the GEB and paid $1 per month by the government "so that they are under obligation to [Rockefeller] for the salaries they receive." He thought farmers would be more likely to trust advice from agents of the government, paid from public funds. "It is natural to expect these demonstrators to realize their obligation to the Standard Oil Company for their salaries,

52. Ibid., (June 25, 1914): H 11133.
53. Ibid.

and it is unwise to have any officials or employees of the Government under obligation to that or any other corporation." Applause followed.[54]

Opposition to the GEB sprang from several issues: the ethics of private influence operating as an arm and within the government, as well as reactions to the more recent Ludlow massacre in Colorado. Eleven thousand coal miners had struck at the Colorado Fuel and Iron Company coal mine, which the Rockefeller family controlled. Violence escalated, and militia called in to force the miners back to work ended up firing machine guns into the striking miners' tent city, which erupted in flames, killing two women and eleven children. Initially, the company and the Rockefellers handled the episode poorly. They blamed the miners for the deaths, which incited a firestorm in the press. John Jr. traveled to Colorado in a public relations effort after the violence, but the mining company's inept handling as well as the Rockefellers' arrogant response aroused the nation. Congressional hearings were concurrently under way regarding the events in Ludlow. New Jersey senator James Martine, a Democrat, reported that he had just come from listening to "sad-faced, wan-faced women, fresh from the mines of Colorado, suffering the brutality and the vengeance of the Rockefeller millionaires." Calling the Rockefeller money "the wages of sin," he added, "If we cannot exterminate the cotton boll weevil without recourse to Rockefeller and Carnegie, then a thousand times rather let it destroy the entire cotton crop, and God in His wisdom and humanity will provide another source for the clothing of mankind."[55]

Senator Thomas Gore, a Democrat and former Populist from Oklahoma, revealed that the GEB held $2.5 million in stock in the Colorado mining companies involved in the "labor war with the miners of Colorado." Democratic Missouri senator James Reed described events in Colorado, noting that the mining interests had imported private police and used political influence to make them state militia, clothing them in the authority of the state, "with the full knowledge that those men, clothed and armed and practically controlled by this corporation, burned a city of tents in which were men, women, and children." He

54. Ibid., H 11135.

55. Raymond B. Fosdick, *Adventure in Giving: The Story of the General Education Board*, 59; Chernow, *Titan*, 578; *Congressional Record*, 63d Cong., 2d sess. (May 23, 1914): S 9121.

argued, "If the money of Rockefeller was used on the boll weevil with the same exterminating effect it had on those little children in Colorado, it would stamp out that pest pretty quickly." Senator Harry Lane, a Democrat from Oregon, argued, "There are many people who believe that there is sweat and blood and grime and mothers' tears on every penny of his money, and that the curse of God rests on the hand that accepts it; that it is used to demoralize the service of the people. They do not want the money, and they do not want the Senate to vote it onto them." It was not farm demonstration work that Congress wanted to end, Lane pointed out; the "prejudice against taking money from Rockefeller" was the real issue. State and local benefactors could fund programs, "but there is an objection to taking money from Mr. Rockefeller and the Standard Oil Company."[56]

As a result of the hearings, the Rockefellers could continue funding whatever philanthropic efforts they liked; they just could no longer do it through the federal government. The demonstration work had been successful, however, and Congress quickly agreed to continue funding the agricultural programs, even expanding them, and that same year the Smith-Lever Act established the Agricultural Extension Service. Eventually, William Spillman appointed four hundred agricultural extension agents, all paid and administered with public funds through state agricultural colleges.

There was no reason that the Rockefeller Foundation could not continue philanthropic efforts for southern farmers after 1914; Congress had simply restricted the GEB from buying influence within the USDA through lavish funding of department programs. Nevertheless, in 1914 the GEB said good-bye to funding southern farm efforts as well as rural education there, explaining that "the South is relatively prosperous, and is willingly devoting steadily increasing funds to school purposes." The GEB did continue funding one person in each state's Department of Education, to push for consolidation of small schools into larger ones, "an improvement likely to be accelerated by the introduction of the inexpensive automobile." Pointing to Louisiana's progress in ridding

56. *Congressional Record*, 63d Cong., 2d sess. (May 23, 1914): S 9122, 9123; (May 4, 1914): S 7682.

itself of one-room schools, the GEB applauded the state for having "300 school wagons in use."[57]

The Smith-Lever Act

It was perhaps no coincidence that as Congress waded through the GEB investigation, another bill was wending its way through the House and Senate that would solve the problem of funding public agricultural extension education. A formal centralized system for directing and disseminating agricultural education to the public—funded by the public—linking land-grant colleges, agricultural experiment stations, vocational high school programs, and the public sector, had been crafted into the Smith-Lever bill. It was the work not of one man, but of many, over nearly a decade of negotiation. Indeed, a variety of bills relating to farm extension education and agricultural vocational education peppered Congress between 1909 and 1914.

In 1911, at the meeting of the Association of American Agricultural Colleges and Experiment Stations (AAACES), the colleges addressed the issue of legislation for agricultural education. At the meeting, Spillman presented a plan for "regional, state, and district field agents, to be financed jointly by the Federal Government and the States," along with a report on farm demonstration work being carried out through the Office of Farm Management in the states where demonstration agent work was occurring (not the GEB or Knapp programs). The AAACES adopted resolutions in favor of an extension bill to support the system Spillman described, stating that "congressional legislation granting aid to the states for this purpose is at the present time of pressing importance for American agriculture and the most approved method of reaching the masses of the people with the best ideals and practices of scientific agriculture."[58]

That same year, the National Soil Fertility League formed, made up of bankers, railroad officials, and businessmen in the Midwest who promoted agriculture. Based in Chicago, members included James J.

---

57. Fosdick, *Adventure in Giving*, 1; *General Education Board: Account of Activities*, 181, 186.

58. True, *History of Extension Work*, 107; Scott, *Reluctant Farmer*, 290.

Hill, Champ Clark, Samuel Gompers, William Jennings Bryan, and former Wisconsin governor William Hoard. They advocated a system of county agents under the direction of land-grant colleges, too, gaining support from President Taft. The AAACES, the National Soil Fertility League, and the Department of Agriculture crafted a bill that was introduced in the Senate in 1912 by Georgia Democrat Hoke Smith, followed by a similar bill introduced in the House by South Carolina Democrat Asbury Lever. A revised bill, termed Smith-Lever, was introduced next; this time, it added the Department of Agriculture as a codirector of the program, along with the land-grant colleges. The final bill included teaching home economics as well as the imperative that the county agent "must give leadership and direction along all lines of rural activity—social, economic, and financial. Not only production but also distribution, must be taught by the extension service."[59]

The bill passed in the House on January 19, 1914, and in the Senate on May 2, 1914, and was signed by President Wilson on May 8, 1914. Wording in the act was altered from the House version approved in January, adding a provision to prevent contributions from large interstate corporations, in particular removing the General Education Board from participation in the work.[60]

The Smith-Lever Act formalized congressional support for the system that was already nearly in place and incorporated colleges into what might have been limited to federal administration. The act created the Cooperative Extension Service associated with each U.S. land-grant institution. It authorized ongoing federal support for extension education, while requiring that the states provide matching funds in order to receive federal funds. Proponents of a federal agricultural education system had been thwarted by the land-grant colleges, which refused to support a federal program because it threatened the research programs on their campuses. Smith-Lever placed the national system of extension agents on the land-grant college campuses, which moved the colleges to support the act. It also provided funds to be matched locally for vocational education at precollege levels, creating vocational education programs in high schools.[61]

59. True, *History of Extension Work,* 112.
60. Ibid., 113.
61. See National Association of State Universities and Land-Grant Colleges, *The Land-Grant Tradition.*

The agricultural colleges had sought funding and control of extension education and had already laid a firm foundation based on farmers' institutes, pamphlets, and traveling lecturers. They were reluctant to adopt either Knapp's demonstration farm system or the agricultural agent idea. They thought a national system of agents would create "a gigantic political machine," historian Grant McConnell explains. Thousands of agents working directly with local interests yet reliant on funding from politicians in Congress promised only to complicate the colleges' roles. It was that intervention by the colleges, however, that McConnell attributes to preventing development of a politically powerful extension service. The colleges had also resisted supporting vocational education in the high schools, feeling it detracted from their mission at the college level, giving students little reason to go on to study agriculture at the college level. By ensuring that high school vocational education teachers would have to be trained at the land-grant colleges, however, they were brought behind the bill. Incorporating home economics, as well as vocational subjects, the plan held wide appeal for the general public.[62]

The Smith-Lever bill caused consternation among the GEB-sponsored programs, too. Seaman Knapp had held a longtime distrust and dislike for colleges and academic men and had done his best to avoid working with them in southern demonstration work. Although it was logical that the expanding extension and demonstration programs would eventually be administered through a central office in each state, and the land-grant college, with its agricultural experiment station, seemed ready-made to run the programs, Knapp refused to work with the colleges. After Knapp's death in 1911, South Carolina, Texas, Georgia, and Florida adopted programs administered by an agent at each state's agricultural college. The other states in the GEB program continued under Bradford Knapp's direction.[63]

As events unfolded in Congress, the Department of Agriculture found itself in an embarrassing position due to the GEB programs. For one thing, the department had to reveal the amounts it was spending in various states, which was problematic because they were, in a sense, running two sets of accounting books because of the GEB funds. Texas,

---

62. McConnell, *Decline of Agrarian Democracy,* 35; Cremin, *Transformation of the School,* 57.
63. Scott, *Reluctant Farmer,* 208.

for example, was receiving $69,136 per year for demonstration work, but the department was only paying $2,200. Mississippi and Alabama each received $43,650 per year, but the department was actually giving them $2,135 and $4,635, respectively. Making the figures public threatened to expose the GEB's influence to states who were not recipients of their programs. To make it even more difficult, the Smith-Lever bill intended to give each state $10,000, four times what some states were receiving yet a cutback to those endowed with GEB largesse. In what may have been an attempt to head off the bill's introduction, in April 1913 a letter from the assistant secretary of agriculture to Senator A. F. Lever, the bill's cosponsor, explained that the proposed large effort at extension work "would require a considerable shifting of funds that are contributed by agencies other than the Government." The department's expenditures for the northern and western programs Spillman was running totaled $168,864. That letter may have been in response to worried benefactors of the GEB program, such as the State College of Agriculture in Georgia, whose president wrote Secretary Houston just after the bill passed the House and Senate in 1914, saying, "I hope the passage of the Smith-Lever bill will in no way disturb the pleasant and harmonious relations existing on account of the withdrawal from the state of any of the funds now being spent here."[64]

Bradford Knapp, Seaman Knapp's son, was another worried constituent. Bradford Knapp was the special agent in charge of farmers' cooperative demonstration work within the Bureau of Plant Industry. As the bill was under debate, he alerted the assistant secretary that Howard Gross, president of the National Soil Fertility League, was traveling through the South, threatening to perhaps stir up trouble over the demonstration work there. "In my territory Mr. Gross would be a thorn in the flesh," Knapp claimed. "He doesn't know anything about our conditions and has never trespassed upon the field but what he has disturbed relationships already satisfactorily arranged." Knapp explained that "there are certain very important and very difficult problems

64. B. T. Galloway to A. F. Lever, April 23, 1913, entry 17, box 87, file "Lever Bill" 170/5/8/4, General Correspondence, Records Group 16, Records of the Office of the Secretary of Agriculture, U.S. Department of Agriculture Archives, National Archives and Records Collection, College Park, MD (hereafter cited as USDA Archives); [Judson?] Soule to Houston, February 17, 1914, box 145, file "Lever Bill" PI 191 E17, 170/5/9/6, ibid.

for the Department to work out in some one or two states, it doesn't seem to me desirable that this man should butt in at this time." He feared "unfortunate complications" if Gross interfered, because "the agricultural colleges are pretty badly mixed up in politics . . . and with the opinions I know he holds, will be very detrimental to the entire plan of fixing up constructive organization under the Lever Bill." Certainly, the Smith-Lever bill promised to complicate the discreet relationship between the GEB and department programs.[65]

Another element of conflict arose out of Smith-Lever: black education. As soon as the bill passed Congress, the National Association for the Advancement of Colored People (NAACP) sent a letter to President Wilson, arguing against the legislation. They opposed the law because it shortchanged black people, who would receive no benefits from the bill. The bill directed that in states where there were two or more agricultural colleges (which was the case after the 1890 Morrill Act funded black and Indian agricultural schools), the state legislature should decide which school got the funds. The NAACP had helped introduce an amendment to the bill that would have ensured equal distribution of the funds, but that amendment failed to pass. The group had to push past Congress and the Department of Agriculture, hoping the president would veto the bill. It was signed by Jane Addams, Herbert Parsons, Moorfield Storey, Oswald Garrison Villard, Joel Spingarn, and William Bennet. President Wilson did not accede to their requests. In fact, Secretary Houston went so far as to contact the governor of Missouri, who thought the funds should be distributed to the state university, the school of mines, and the Lincoln Institute (an agricultural school for blacks), telling the governor that to avoid waste, the funds should be given to one institution—the state college—and black farmers who "might be worthy" could be invited in from the Lincoln Institute for programs. The matter did not end there, as in August, Booker T. Washington, from the Tuskegee Normal and Industrial Institute, wrote to Houston, reminding him that when the department delegated funding under the bill, black education in the South should be treated fairly. "I very much fear that in most of the states, the Negro will get very little if anything from this fund," Washington wrote. "I am

65. Knapp to Galloway, March 12, 1914, box 145, file "Lever Bill" PI 191 E17, 170/5/9/6, ibid.

sure you will agree with me in this; no strong race can treat a weaker race unjustly without the stronger race being hindered as much or more than the weaker race." He urged Houston to do the right thing, adding, "I am also sure you will agree with me that it is impossible in a large measure to do any effective and practical work for the colored farmers except as it is done through and by colored people themselves." The conflict over black equality was finally resolved by setting up two separate branches, the Office of Extension Work, North and West, and the Office of Extension Work, South.[66]

"The controversy over the bill was long and bitter," according to Murray Benedict. Farmers thought it was geared to increase production, which they did not want; rather, they wanted better markets. Others claimed it was socialism or class legislation. Despite conflict among nearly every stakeholder group in the country, the bill became law on May 8, 1914, and the Cooperative Extension Service formed as a coalition between federal and state agricultural colleges. Each state received an annual allotment of ten thousand dollars, plus an initial start-up fund of six hundred thousand dollars. Funds would increase each year but would have to be met by state sources. New programs evolved, and existing ones were realigned. In agricultural education, extension education provided by agricultural colleges and universities through county extension agents prevailed. Alternatives faded. Chautauquas continued for a time, and Spillman frequently lectured on the Chautauqua circuit along with other agricultural educators such as Liberty Hyde Bailey. But agricultural education had moved from popular local efforts to an organized national system directed by the Department of Agriculture and administered through state colleges. The land-grant institutions opened the way to federal funding of public research and instruction, according to historian Vernon Carstensen. He explains, "The administrators of the land-grant colleges found a way of obtaining federal funds without yielding entirely to the domination of the federal bureaucracy."[67]

The relationship among land-grant colleges, farmers, and professionals had one thing in common: the application of science to rural

66. NAACP to Wilson, May 4, 1914, ibid; Houston to David R. Francis, June 25, 1914, ibid; Washington to Houston, August 4, 1914, ibid; Benedict, *Farm Policies of the United States, 1790–1950: A Study of Their Origins and Development,* 5.

67. Benedict, *Farm Policies,* 153; Carstensen, "A Century of the Land-Grant Colleges," 37.

problems. "The application of science to agriculture was both a scholarly act and a political one and each was democratizing," historian George McDowell contends. "In addition to the informal education role that extension was established to play in response to the political demands of agricultural interests, it is also a major political arm of the system, collecting grass roots support from the clients it serves." The land-grant colleges had a social contract, to serve the public interest and provide democratic educational opportunities, and extension education served that purpose. The USDA and the land-grant colleges were established the same year, 1862, and their relationship in the early twentieth century was "one of rather disjointed and separate interests," according to McDowell. Over the century, the relationship between the "people's universities" and the "people's department" (a term Abraham Lincoln coined) has changed, but in the Smith-Lever Act a nonpoliticized formal working relationship between both was established that has gone unchanged since its inception.[68]

Secretary Houston reorganized the department following the passage of Smith-Lever, moving all extension work to the new Office of Extension Work, under the direction of the new States Relation Service, which directed both the Office of Experiment Stations and the Office of Extension Work. The States Relation Service maintained two offices, one for agents in the North and West and another for the former Knapp region in the South. Longtime loyalties and ideology died hard. As farm practice work became the domain of the county extension agent, the work of the Office of Farm Management shifted away from studying farm practice and began to emphasize cost accounting and agricultural economics. The Smith-Lever Act was both a culmination and a turning point in agricultural education. As Roy Scott explains, "For more than a century men had sought methods for educating the mass of farmers; by trial and error they had now found it."[69]

Spillman and the agricultural colleges had agreed (some reluctantly) to the Office of Farm Management's involvement in setting up a national system of agricultural agents operating through agricultural colleges. Salaries came from state legislatures, business donations, federal government, Granges, and local schools. The agents were free to

68. McDowell, *Land-Grant Universities and Extension into the 21st Century*, 127.
69. Scott, *Reluctant Farmer*, 313.

develop programs they saw fit for local needs. Through a national system of local agents, Spillman linked the scientific agriculture being done at the colleges and experiment stations to the general public. As Roy Scott says, Spillman "was fusing the knowledge of the various bureaus of the Department of Agriculture and the agricultural colleges with the proven practices of successful farmers to provide a new kind of comprehensive data useful in the proper operation of any given farm anywhere in the United States."[70]

70. Ibid., 260.

3

Science, Politics, and the USDA

B y 1915, the Department of Agriculture had increased in size and scope exponentially from its earlier status. It was evolving into a modern bureaucratic organization, subject to public, political, and corporate pressures as well as the personalities and abilities of the bureaucrats directing it from within. Spillman, as a second-tier manager at the department, in charge of the Office of Farm Management between 1905 and 1918, found his role both satisfying and challenging. When James Wilson left the department in 1913, replaced by David Houston, the dynamics within the department changed—enough that the public and Congress began to take a critical look at the department's expanded scope as well as budget, demanding practical, if at least visible, results.[1]

Houston took charge of a department that had been under continuous leadership by James Wilson for three presidential administrations. Wilson had brought the department from the nineteenth century into the twentieth, in many ways. He had expanded the scope of the department, from a staff of 2,444 in 1897 to 13,858 in 1912 and a budget during the same period that increased from $3.6 million to $21.1 million. An astute administrator, Wilson appointed men by merit or ability and then gave them freedom to pursue solutions to problems in agriculture. Houston, however, saw the need to change the dynamics of the department through a restructuring of its many offices. He consulted first with Walter Hines Page, Wallace Buttrick and F. T. Gates (both of the

---

1. See A. Hunter Dupree, *Science in the Federal Government: A History of Policies and Activities to 1940;* and Eva Etzioni-Halevy, *Bureaucracy and Democracy: A Political Dilemma.*

General Education Board), and Thomas Carver, professor of economics at Harvard. Beverly Thomas Galloway, former chief of the Bureau of Plant Industry, served as his assistant secretary. Galloway was valuable because he had been with the department since 1887 and knew how things worked; he did not last a year under Houston before taking a position as dean at the New York State College of Agriculture at Cornell, replacing Liberty Hyde Bailey.[2]

Houston's reorganization created departments based on functions, rather than bureaus. He claimed it was necessary because work was uncoordinated and because regulatory work did not mesh in the same office with research work. His goals were to increase efficiency by simplifying operations. As a result, in 1915, the Office of Farm Management lost several areas it had formerly administered: the farm home management and extension agent work was moved to a new office, States Relation Service; farm credit and insurance work went to the Office of Markets and Rural Organization; farm architectural work went to the new Office of Public Roads and Rural Engineering; and the investigations of dry land plants and weed eradication were shifted to the Bureau of Plant Industry. With Spillman's authority diced up and divided among several divisions, his Office of Farm Management was then removed from the Bureau of Plant Industry and put directly under the direction of the office of the secretary of agriculture. The only division so managed, it was clear that Spillman was losing not only his investigational fields but his authority and independence as well.[3]

Houston concentrated more power in the office of secretary than Wilson had ever envisioned, as he centralized operations and became the clear authority for policy matters. He gave additional authority to the position of assistant secretary, a presidential appointment held by Clarence Ousley, who had been editor of the *Fort Worth Record*. Houston ordered that the office of secretary would now oversee all requests for new experiment stations and test farms; all correspondence on exhibits, fairs, and the like; and all requests for information on appropriations. All new projects or lines of work were to be approved by the assistant secretary before work was begun. A committee would review all ongoing

2. Rasmussen and Baker, *The Department of Agriculture,* 14; *American National Biography,* s.v. "Galloway, Beverly Thomas."
3. Baker et al., *Century of Service,* 66.

projects and advise the secretary on their merits. The Office of Inspection was created to examine all department expenditures and to supervise the conduct of all employees (both professional and clerical). The department was clearly becoming a different place to work than it had been.[4]

In 1917, the popular farm magazine *Country Gentleman* ran a series of articles critical of the Department of Agriculture titled "What's the Matter with the Department?" The articles focused on several areas of concern that were dividing the department, including employee morale, increased bureaucratization, and censorship of research publications. Spillman, chief of the Office of Farm Management, was directly at odds with departmental leadership over several issues, essentially those identified by the magazine. The situation became so untenable that he resigned from the Office of Farm Management in 1918.

Waning employee morale was rooted in the increasing bureaucratization of the department and exacerbated by centralized control under David Houston's leadership. Censorship of publications became increasingly common as bureaucrats allied with industry, negating an open flow of information to the public.

Background at the Department

The effectiveness and success of the department depended upon the secretary and his staff. James Wilson had served from 1897 to 1913 as secretary of agriculture, the longest continuous term of any cabinet officer in U.S. history. During his tenure, civil service reform became a reality as the Pendleton Act of 1892 was put into effect; it both helped and hindered his attempts to select the best personnel. Each cabinet member could appoint a private secretary, but other employees had to pass through a screening system based upon an examination for the job. "Civil Service is operating completely," Wilson noted. The system saved him time and made the office more efficient, but it limited his autonomy. He believed "the doing of some few things that perhaps would be better done without Civil Service" was stifled. Wilson found it stultifying when he wanted to give jobs to friends of political col-

4. Ibid., 69.

leagues. He tested the system by hiring Union army veterans without examination and promoted employees the previous secretary had demoted. The department hired large numbers of laborers, both clerical and agricultural, positions Wilson gladly filled with names sent by congressmen. Even Francis Scott Key's granddaughter was a "laborer" on departmental rolls during Wilson's tenure. His private secretary, the only staff position he could freely assign, went to his sons, first James Jr., then Jasper, who resigned in 1911 after a scandal.[5]

Wilson has been lauded for the team he assembled during his long tenure in the department. He brought a lively assortment of scientists, educators, and agricultural writers together, then gave them free rein to design and institute programs. Gifford Pinchot once said, "James Wilson did more for me than any other man on earth. For twelve years I had the greatest time of my life." Nevertheless, the situation gradually became less flexible, and Wilson was constrained in selecting personnel as the system became more rigid. He kept the heads of the bureaus and divisions as appointments, rather than part of the classified service, and because he began to promote from within, problems seldom arose.[6]

James Wilson's replacement in 1913 was David Houston, who brought an entirely different leadership personality to the position, as Woodrow Wilson installed the first Democratic administration since 1896. By that time, the civil service system was well entrenched, and a change in presidential administrations merely replaced the cabinet officers and their assistants, leaving the body of each department to continue under leadership that was at times going in a totally different direction.

Beginning in 1905, a series of scandals in the department made news. The worst was the revelation that crop estimates were secretly leaked before their official release date, allowing speculators to profit. Cotton growers uncovered the scheme in which Edwin Holmes, a statistician at the department, used his window shade to signal crop speculators on the street below. It had gone on for years, and Holmes and his friends made big profits from the commodities markets. Wilson suspended Holmes but did not fire him. He tried to deflect attention to the speculators and not the department. "Although the press also accused Wilson

5. Willard Lee Hoing, "James Wilson as Secretary of Agriculture, 1897–1913," 26–27.
6. Wilcox, *Tama Jim*, 70; Hoing, "Wilson as Secretary," 30.

and his son Jasper, who was his personal secretary, of being involved in the scandal, there was never any evidence to support the accusations," explain historians Clayton Coppin and Jack High. John Hyde, chief of the Bureau of Statistics, resigned, and James Wilson and son Jasper exited Washington for a tour of western forests until things calmed down.[7]

There were plenty of other scandals to tarnish Wilson's reputation, however. The chief of the Weather Bureau was spending lavishly beyond his means; Daniel Salmon, chief of the Bureau of Animal Industry, was forced to resign due to profiteering from the company that printed meat inspection labels. In 1905, a congressional investigation found Wilson unable to control the department. Wilson, however, a skilled entrepreneurial bureaucrat, was popular with farmers and used that influence to retain his position.

President Theodore Roosevelt and James Wilson wanted to fire Harvey Wiley, chief of the Bureau of Chemistry, but were stymied by his popularity with the public as well as his political connections. Roosevelt told one of Wiley's critics that "Dr. Wiley has the grandest political machine in the country." Wiley's huge popular following, particularly from the growing consumer movement, annoyed Wilson. At one point, he remarked that "the short-haired women" were out to defend Wiley "who spends considerable time talking to them." Later, President William Howard Taft found Wilson's management of the department deplorable. "The Wiley business is a mess and I am inclined to think I may have to get a new secretary of agriculture," Taft lamented, "but he stands well with the farmers and it might be difficult to get rid of him."[8]

Wiley and Wilson had both kept their positions and held their bureaucratic ground by taking their issues to the public, Wilson to farmers, Wiley to consumers. On March 15, 1912, however, Wiley resigned from the Bureau of Chemistry. No longer willing to put up with the frustrations he encountered from administrators, he left the division he had almost single-handedly created.[9]

In 1912, two weeks after Wiley's resignation from the department, Wilson was under scrutiny again. The Moss Committee investigation

7. Coppin and High, *The Politics of Purity: Harvey Washington Wiley and the Origins of Federal Food Policy*, 84.

8. Ibid., 87; Hoing, "Wilson as Secretary," 258, 257.

9. Wiley, *The History of a Crime against the Food Law*, xi.

in Congress questioned Wilson's role in suppressing a report on the drainage of the Everglades to the benefit of promoters, as well as department involvement in financial problems that led to a grand jury investigation and possible prison sentences for three employees. Congress attempted two other investigations of Wilson and the USDA, one connected with meat packers' violations of the Meat Inspection Act of 1906 and the other questioning Jasper Wilson's role in federally funded dam projects in Colorado. Speeches on the floor of Congress grew so heated that two were erased from the permanent record by action of Congress. Although no punitive actions resulted, the issues resonated through the popular press, adding to the controversy surrounding the aging Wilson's tenure as secretary of a burgeoning and more influential Department of Agriculture.[10]

The *Country Gentleman* series asserted that the department's policies had become too bureaucratic and that it was "drifting away from the viewpoint of the farmer." Journalist William Harper Dean, citing a litany of complaints, criticized the department for ignoring work that was crucial to farmers and business, such as the national fertilizer shortage and the need to develop additional potash sources. The crop statistics reports were given out with no interpretation of the data for farmers because the department claimed that to make predictions or provide advice was against policy. Dean referred to a common saying that "the Weather Bureau is the only branch of the Department of Agriculture that is allowed to forecast." The bureaucracy and attitude of the department had become inefficient, Dean said. Yet he pointed to Spillman's Office of Farm Management as a model exception. He stated that William Spillman, chief of the Office of Farm Management, was responsible for a broad range of programs, "yet this office looks ahead, plans its work with an end in sight, and shapes its requests for appropriations

---

10. Hoing, "Wilson as Secretary," 267–68; Christopher F. Meindl, Derek H. Alderman, and Peter Waylen, "On the Importance of Environmental Claims-Making: The Role of James O. Wright in Promoting the Drainage of Florida's Everglades in the Early Twentieth Century"; H. Parker Willis, "Secretary Wilson's Record"; U.S. House of Representatives, *Hearings before the Committee on Expenditures in the Department of Agriculture: Florida Everglades Investigation, Congressional Record,* 62d Cong., 2d sess., 1912.

accordingly." Clearly, Spillman and the Office of Farm Management enjoyed solid support from the farm press, at a time when the rest of the department was a focus of concern.[11]

Indeed, power struggles between bureaucrats dominated the Department of Agriculture during the early twentieth century, as it became a hotbed of jealousy and politicking. William Harper Dean, author of the *Country Gentleman* series, claimed that "bureau chiefs in the Department are constantly on guard lest some other bureau pounce down and snatch a line of work away from them." Entrepreneurial bureaucrats, they sought to jockey for power, authority, and budgets in order to expand their divisions while diminishing the opportunities for rivals within the bureaucracy.[12]

Censorship of Publications

Public concern about political influence within the department centered on how that influence shortchanged farmers. The "torturous channels to publication" that operated within the department robbed the public of much information that could be useful on the farm. Research on things like fertilizer chemistry was invaluable to both farmer and industry, but the department held back reports from publication. Phosphate prices soared, but the department did not alleviate the situation with valuable facts about alternatives, according to Dean.[13]

Dean described a frustrating experience he had regarding a report on the cost of producing oranges, which presented orange growing as having less than glowing potential. The information would have dissuaded investors from a region then under development by land promoters. Dean was advised by a department employee to "keep off that report" because printing the findings in *Country Gentleman* would "make trouble." Dean kept digging and discovered that the Office of Farm Management had received a memo from "higher up" ordering the study to be discontinued.[14]

---

11. Dean, "What's the Matter with the Department?" (March 3, 1917): 415, 416; Dean, "What's the Matter with the Department?" (March 17, 1917): 8.
12. Dean, "What's the Matter with the Department? The Confusion of Tongues"; Etzioni-Halevy, *Bureaucracy and Democracy*, 84.
13. Dean, "What's the Matter with the Department?" (March 17, 1917): 8.
14. Ibid., 9.

Whether the work was stopped due to political reasons was not Dean's issue; it was the wasted effort—another example of expensive inefficiency rampant at the department. Dean described the waste inherent in the publications office, including large numbers of research publications were never published and some that were published were so mundane (or edited to be so) they were laughable.

An examination of the records of the Office of Farm Management during this period reveals that Dean was at least partly correct. Several studies that Spillman's subordinates sent for publication were rejected by the department's administration, and his own critiques were questioned. Spillman attempted to publish a number of articles that were discouraged by his superiors. Publications from the Office of Farm Management had to be approved by his superiors at the Bureau of Plant Industry and their superiors in the secretary of agriculture's office. In 1913, a department employee had invented a fertilizer-processing machine that Spillman thought the department should patent and make available to manufacturers, but L. C. Corbett, assistant chief of the bureau, discouraged the idea, claiming that manufacturers would not be interested if none could have a monopoly on the machine. (Perhaps this is a reference to the alternative to increasingly expensive fertilizers Dean wrote about.) Corbett also criticized Spillman's views on tenant farming in an article he was submitting to the *Breeder's Gazette* about tenant farming and land values. Corbett was concerned that Spillman's views were not conservative enough and believed that Spillman should not criticize tenant farming; in fact, he saw it as the future of agriculture and encouraged Spillman to find ways to approach it positively. "It seems to me that we should . . . perhaps inaugurate a movement which would have for its object the betterment of tenant farming rather than the discouragement of it," Corbett told Spillman. "I feel that tenant farming is with us to stay, that it will increase rather than decrease, and that what we need is better tenant farming rather than anything else." Corbett supported his stance by pointing to England, where tenant farming was well entrenched, and took a Malthusian stance, claiming, "I am convinced that the tenant farmer will be very much more numerous in the future than he has been in the past. It is the same old question of the poor, simply stated in different terms." Criticisms about the growing number of landless farmers, especially in the South, would not appear in department literature. Instead, the ideal of the "agricultural ladder" was touted, which claimed tenancy was a transitional step on the hierarchy of opportunity.[15]

Department publications were also careful to avoid any suggestion that farmers should limit production. William A. Taylor, chief of the Bureau of Plant Industry and Corbett's superior, criticized Spillman's ideas about the relationship between overproduction and farm income, pushing Spillman to eliminate any wording that might question the idea of full production on U.S. farms. "I think it would be wise," Taylor explained, "to eliminate the suggestion even of limiting the production in an article of this sort.... [T]o suggest the possible diminution of the aggregate supply through cooperative efforts or through educational efforts I think would better be omitted." Spillman argued that he wanted to "convey the idea that the great hue and cry for increased yields will not solve the farmers' problems and might even be ruinous to the farmer" and that "increased output should merely keep up with increased demand and not exceed it." Regarding another article, "The Tendencies and Dangers Involved in Existing Land Values in the United States," Corbett warned Spillman that "for obvious reasons I think it is well not to use any phraseology which could be twisted into an interpretation that we meant to imply that by some method of curtailing production, prices to the farmer could be forced up."[16]

Taylor also held back publication of an article by some of Spillman's employees regarding the difficulty small farms were having making a profit in Oregon. Spillman sent the paper on to be vetted by Taylor, with the hopes that "the publication of these results will be of distinct value to a large number of people who are buying small farms in the north-western states." Taylor questioned the authors' conclusions and thought that the high land value ($205 per acre) was less important than the "uneconomic size of the farming unit, improper selection of crops, or the incompetency of the operators." Publication plans for the paper were tabled while data were assembled on the profitability of larger farms, and Taylor noted to Spillman that it should be printed as

15. Corbett to Spillman, October 17, October 9, 1913, box 101, Records Group 54, no. 49, Bureau of Plant Industry "Bureau Chief's Correspondence, 1908–1939," USDA Archives.

16. Spillman to Corbett, October 15, 1913, ibid; Corbett to Spillman, October 18, 1913, ibid.

a departmental bulletin anyway, rather than for the Farmers' Bulletin series.[17]

In another Oregon case, Spillman questioned information in the *Reclamation Record,* a publication of the Department of the Interior, with cooperation from the USDA's Office of Demonstrations on Reclamation Projects. It touted farms located on recently irrigated land near Umatilla, Oregon, claiming the farmers were "quite well-established and apparently have their places close to an income basis." Spillman told the chief of the Bureau of Plant Industry, which both oversaw the reclamation demonstration farms and published the *Reclamation Record,* that the farmers there had wasted several years of effort in a fruitless quest. "Many of the settlers on this sandy hill tract have had to abandon their claims after spending all they had on improvements and in an effort to make their land produce enough to pay expense," he explained. "The Reclamation Service people well know that most of the soils in this project are exceedingly infertile." Calling the information "a gross misrepresentation of fact," Spillman had been alerted to the article by extension agents working in eastern Washington, who viewed the reclamation project with concern. Agricultural experts had too often been dragged into boosterism for land development schemes, and as the department's agriculturalist in charge of farm management investigations, Spillman argued that the department should counter the misinformation. Taylor replied, "While the wisdom of publishing the article referred to may well be questioned, it does not appear to me that it is likely to do any particular harm . . . and its non-official character is quite obvious, even though it was printed in the *Reclamation Record.*"[18]

Criticisms of agricultural industrialization were carefully screened in publications that went out to the public. In 1915, a manuscript by Spillman's employees, "Farm Experience with the Tractor," was rejected by the department's Committee on Examination of Manuscripts because the "general tone of the manuscript was destructive rather than constructive," according to Taylor. "Even though the facts may justify this character of expression to a considerable extent, it would be wiser to

17. Spillman to Taylor, August 13, 1913, ibid; Taylor to Spillman, August 28, 1913, ibid.
18. Byron Hunter to Spillman, May 25, 1914, ibid; Spillman to Taylor, June 3, August 20, August 25, 1914, ibid; Taylor to Spillman, September 3, 1914, ibid.

point out more conspicuously the way in which some of the difficulties can be met." The authors attempted to approach tractors as objectively as possible, noting that most information had been biased and little actual information about performance and economics had been published. Their investigations revealed that tractor plowing could not be done as deep as with horse-drawn implements, that cost of repairs increased by seven times between fifteen- and forty-horsepower tractors, and that smaller tractors were more successful than larger machines. Too, tractors had a short life span, at most six years, but frequently three years. Most farmers had not replaced horses, relying on them for precision work like drilling seed, so tractors merely added to overall costs. The study found that tractors were an unprofitable investment for most farmers and that many had abandoned them because they did not meet expectations. Ultimately, because tractors had not replaced horses, farmers were spending more capital on power than they had previously, and the tractor had "proved to be an auxiliary of the farm horse rather than a substitute."[19]

Houston's assistant secretary, Clarence Ousley, insisted on inserting changes to the manuscript, smoothing the negative tone of the study before the tractor bulletin's publication in 1915. It was so popular that the supply was exhausted a few days after publication. Three thousand reprints were ordered with only minor changes, but the bulletin did not please everyone. One critic lambasted the department in *Farm, Stock, and Home,* claiming the "data was gray-haired before it got to the printers" and that "it is especially unfortunate that a bulletin purporting to give reliable data on modern tractors should have in it practically nothing up to date." The writer blamed it on "red tape" and thought "the temptation is strong to go after the offending bulletin and its writer" but that a "pigeon-hole cleaning and big bonfire" would suffice.[20]

Ultimately, the study's conclusions were apt: U.S. farmers needed a smaller, more versatile tractor, not larger, inefficient models. Within days

19. Arnold P. Yerkes and H. H. Mowry, *Farm Experience with the Tractor,* 1, 43; Wayne D. Rasmussen, ed., *Readings in the History of American Agriculture,* 160; Robert E. Ankli, "Horses vs. Tractors on the Corn Belt," 146; Spillman, *Farm Science: A Foundation Textbook on Agriculture* (1918), 290.

20. "A Chronic Trouble," *Farm, Stock, and Home,* undated clipping, 1915, box 102, Records Group 54, no. 49, Bureau of Plant Industry "Bureau Chief's Correspondence, 1908–1939," USDA Archives.

of the bulletin's first publication, Henry Ford purchased two thousand acres along the Rouge River in Michigan and embarked on manufacturing a small economical tractor, the Fordson, which was exactly what Spillman and his colleagues had in mind. In the year the Fordson was introduced (1917), it took 24 percent of the tractor market, and by 1920 one-third of all tractor sales were Fordsons. In 1924, International Harvester introduced the Farmall, a competing model that included a power takeoff, making implements more efficient. Small tractors had triumphed.[21]

During the Progressive Era, the press was a significant factor in American life, particularly the agricultural press. Professionals in the department published information through official outlets, but many also wrote for the agricultural newspapers and magazines. Secretary of Agriculture James Wilson wrote for newspapers and magazines and edited a newspaper before coming to the department. He was closely connected with William Dempster Hoard, publisher of the largest farm magazine during that era, *Hoard's Dairyman,* as well as Henry Wallace Sr., founder of *Wallace's Farmer.* Three secretaries of agriculture came to their position as publishers of farm magazines: Edwin Meredith (1920–1921) published *Better Homes and Gardens* and *Successful Farming,* and Henry C. Wallace (1921–1924) and Henry A. Wallace (1933–1940) published *Wallace's Farmer.*[22]

At the dawn of the twentieth century, two major shifts of emphasis were occurring within the department: the emphasis on science and the expansion into the popular press. Agricultural journalism slanted toward the public interest had been ignored by the media during James Wilson's tenure; the press paid scant attention to the Department of Agriculture, except to cover political scandals. The press covered the conflicts within the department extensively, from Harvey Wiley's 1907 food and drug legislation controversy to the crop-statistics scandal, Florida swamp drainage schemes, and the General Education Board's influence on the department. As the Progressive Era matured, however, the press became vital to building coalitions between bureaucrats and the

21. Robert Lacey, *Ford: The Men and the Machine,* 470; Robert C. Williams, *Fordson, Farmall, and Poppin' Johnny: A History of the Farm Tractor and Its Impact on America;* Reynold M. Wik, *Henry Ford and Grass-Roots America;* Ankli, "Horses vs. Tractors," 135.
22. Hoing, "Wilson as Secretary," 10; Wilcox, *Tama Jim.*

public, providing an information conduit for the public as well as validation for government agencies. The press was a double-edged sword, however, giving prominence to the department but also allowing a venue for airing claims and conflicts within and concerning the department. Governmental focus on propaganda and the press during World War I demonstrated the importance of the press in swaying public opinion. Not only did the department expand its own use of the media, sending out press releases and publications to gain public support, but well-known professionals at the department such as Harvey Wiley, Gifford Pinchot, and William Spillman also frequently published articles in popular magazines. Spillman served as editor at the *American Naturalist* while he worked at the department. Wiley, too, used public print both to spread his opinions and to bolster support for his program at the department. After Wiley left the department in 1912, he took a position at *Good Housekeeping*, where he continued to advocate for consumer interests.[23]

Spillman Out

In 1918, the Office of Farm Management was investigating two major issues in agriculture: land classification and land tenure. Spillman was administering a study to determine exactly how much potential farmland was available, information that would give experts and farmers a better foundation for planning. Spillman grounded the study in the same census data Frederick Jackson Turner studied and noted that during the decade ending in 1910, farmland prices had increased 108 percent, far outdistancing the rise in farm prices. Connected with land status, a second study looked at landownership and how farmers acquired ownership status compared to tenancy. He believed that a study of tenancy, the relationship between land prices and rental values, and the length of time for tenants to acquire ownership status were important. The lease contract was under examination, with men in his office collecting thousands of contracts and analyzing the relationship between actual incomes of landlord and tenant.[24]

23. Etzioni-Halevy, *Bureaucracy and Democracy*, 84, 98.
24. Spillman, "Work of the Office of Farm Management Relating to Land Classification and Land Tenure" (1918), 65, 70.

Along with farmland, he wanted to examine ranch ownership practices and the relationship between numbers of animals and land use, noting that graziers using public lands could better conserve range feed if they could keep competitors from exploiting range they tried to conserve. Spillman noted that research showed the number of animals grazing on public lands could be increased through control of rangeland, rather than blatant competition and exploitation between ranchers. "It may therefore be assumed," he wrote, "that when a proper basis has been determined for legislation on the subject and the range, by some method of leasing, homesteading or otherwise, has been parceled out among ranchmen so that each has exclusive control of his own range, a material increase will result in the amount of stock produced in the range country."[25] Public-land grazing would later be controlled with the passage of the Taylor Grazing Act in 1934.

Spillman's ideas were grounded in applying scientific agriculture to practical problems, countered by his own social conscience. He was articulate, popular, and personable. Yet he did not get on well with Secretary Houston. Spillman, once highly touted as a potential nominee for the secretary of agriculture post, was continually at odds with David Houston's administration of the department. Harry McClure, an employee who worked under Spillman in the Office of Farm Management, described the situation: "What a struggle Spillman had in bringing into existence the new Science of Farm Management and how bitterly the 'whole department' fought him and his men when instead of trying to destroy the new work they should have been proud of it." McClure described the working environment under a punitive Houston: "The old Office of Farm Management was broken up by methods that should never be allowed to be used in any civilized country. Many were put in 'solitary confinement,' i.e., made to sit at a desk in idleness for days, weeks, and months, in order to break their spirit and force them to resign, which some did. Others stayed on and to this day not one of Spillman's competent men has ever gotten a fair deal." McClure wrote, "I was transferred to [the Office of] Markets and produced the first fair play and honest grades for a farm crop—hay—in the history of the Dept." McClure described a controversy in the department about methods for artificially drying hay, a concept that would have allowed

25. Ibid., 71.

more southern farmers to expand into livestock rather than cotton. The bulletins were ready for printing when Houston appointed Henry Taylor to replace Spillman, and they were immediately "killed" because, as Taylor later explained to McClure, "I wanted to destroy everything done by Spillman in the Office of Farm Management and as you had gone farther in a line of research than any of Spillman's men, yours was the first work to be stopped." McClure had come to the department in 1906 after graduating with a master's degree in agriculture from Iowa State Agricultural College in Ames and had worked in the department under Spillman until Spillman's resignation in 1918. McClure noted, "Spillman was the fairest and squarest man I ever knew and entirely too brilliant a man for Gov't Service."[26]

In June 1918, frustrated by continued resistance to publication of his ideas, Spillman quit the department and in October joined the staff at *Farm Journal* as an editor. He had been offered the position of dean of agriculture at Washington State Agricultural College, but he turned it down, saying he could be more effective at the magazine in his efforts to reach the public regarding agricultural policy. It was a significant time for the agricultural press, and Spillman saw it as an outlet through which he could communicate his ideas about agriculture and economics that had been stifled under David Houston's tenure as secretary of the department.

On a light note, the present Spillman's colleagues presented him with at a going-away party reflects the dynamism and creativity of those who worked for him. His papers contain a well-preserved copy of a pseudopublication the Office of Farm Management staff prepared as a joke for Spillman's departure. Called "Farm Management Investigations as Applied to the Measurement of Terrestrial and Sidereal Time in Accordance with the Principles of Economics and Psychology: An Epileptic Farm Survey in Five Fits," it was an "EMERGENCY BULLETIN" replete with a replica of Spillman's signature. A copy of the document appears in Appendix B. It claimed to be compiled from "unwritten remarks" and "unspoken thoughts" and was "officially read before Professor Spillman on the occasion of his regrettable yielding to the siren of agricultural journalism August 31, 1918." Satirizing censorship in the

26. McClure to Ramsay Spillman, February 14, 1933, box 6, file 53, Spillman Papers, MASC.

office, the authors joked, "The manuscript shows something of the scintillation that may be expected from the Office of Farm Management
workers when they are relieved of the incubus of a publication committee," adding "this manuscript has not been censored." It continued,
"Hence the springiness, life, personal touch, intimate revelations and
bold treatment." The rest of the bulletin is a humorous parody of departmental bulletins; poking fun at themselves, the authors examined
the use and care of the Ingersoll watch. "The Cost of Time" parodied
statistical analysis of farm practices, "Classes of Farmers and the Ingersoll Habit" poked fun at statistical classification, and, finally, "Genetics
of the Ingersoll Watch" focused on Spillman's genetic achievements. It
was elegantly done, poking fun at "[M]endelism in the ownership of
Ingersolls," pointing out that the "passion to possess Ingersolls appears
to be a clear case of sex-linked character being inherited only by males."
The final punch line said: "A questionnaire sent to the 76 farmers failed
to bring in any important additional information. Seventy-five replied
that the Government had no right to make such personal inquiries
while one farmer frankly stated that he refused to be the heterozygote
in this matter." The term *heterozygote* refers to an organism carrying
different alleles on connecting chromosomes—an "inside" joke directed
at Spillman's genetic work.[27]

The "emergency bulletin" must have been very special to W. J. Spillman, not just because his colleagues had clearly identified and parodied what he had achieved at the Department of Agriculture but also
because it showed their clear admiration and affection for him. Leaving
the department must have been difficult, but the situation had deterio-
rated perceptibly, particularly for Spillman.

Spillman published extensively in the *Farm Journal* both under his
own name and in anonymous articles, as well as in several other venues
over the next few years, notably an agricultural textbook, a variety of
other rural magazines, and professional journals. One theme resurfaced
in his writings again and again: the idea of land tenure and the growing
number of landless tenant farmers. He took his message about tenancy
to the American Economic Association, where he spoke to the annual

27. "Farm Management Investigations," Spillman Papers, MASC. Copy also in
Special Collections, National Library of Agriculture, U.S. Department of Agriculture,
Beltsville, MD.

meeting, pointing out the agricultural ladder and its shortcomings in the face of growing farm tenancy. At the bottom rung was the farm boy who worked on the home farm, the next step was the hired man, the third a tenant, and finally the farm owner. At each step, income and security should increase, and opportunity to move up should be natural. This version of the American dream had roots in the Jeffersonian-Jacksonian ideology of yeoman farmers, but things had changed, as Spillman pointed out. Higher land prices and lack of credit meant that many farmers had to remain tenants, and an increasing proportion of landownership related to inheritance or marriage. At the time, 62 percent of farmers in Illinois acquired farms through kinship or marriage; only 38 percent purchased from others. The "hired man" role was natural to most farmers who saw it as a stage, somewhat akin to apprenticeship, on the way to owning their own place. But that had changed due to the rising price of land. The length of time men worked as hired hands or tenants increased, and farm labor was no longer moving young men up to ownership. He believed that if inheritance and marrying the farmer's daughter were more significant than the touted agricultural ladder, the government should intervene.[28]

Spillman compared the situation to Europe, where land prices related to rental income, with twenty-five years' rent considered the normal price for farmland. The governments in European nations were assisting young men in acquiring their own farms. "There is no fundamental reason why this country can not do the same thing," Spillman pointed out. He urged economists "to consider carefully the entire subject of tenant farming in this country with a view to seeing that it occupies its proper status in a system in which ownership farming is the rule." He recommended that the members of the American Economic Association lobby for legislation to help young men move up from tenancy to landownership. "In helping tenants to buy farms it would be legitimate to limit the purchase price," he suggested, "to a specified number of years' rent." This would cap inflation of land values and reduce tenancy to normal levels.[29] He was careful to not argue completely against ten-

28. Spillman, "Work of the Office of Farm Management Relating to Land Classification and Land Tenure" (1918), 70.

29. Spillman, "The Agricultural Ladder" (1919), 178; Keijiro Otsuka, Hiroyuki Chuma, and Yujiro Hayami, "Land and Labor Contracts in Agrarian Economies: Theories and Facts."

ancy and did not advocate giving credit to men who had not shown themselves to be able farm managers through tenure as a hired hand or tenant. Yet the idea of controlling land values was far ahead of its time, and the implications for capitalists were serious.

Land tenure was a controversial issue in the 1920s, as socialism cast a shadow over a growing disparity of landownership. In 1880, 25 percent of farmers were tenants; by 1900, that number had risen to 35.5 percent and by 1910 was at 37 percent. A discussion at the annual meeting of the American Economic Association in 1919 reflected concerns that the situation was leading "administrative officials and the public into grievous errors" due to reformers' belief that farmers should own the land they worked. This movement opposed the idea of investors who did not actively participate in farming the land they owned. Although governmental intervention and management of farmland ownership might dislodge nonworking investors, it would not create the opportunity the system allowed for tenants to accumulate capital by farming leased land. Spillman explained that "young men starting out with little capital frequently find it financially desirable to be tenants rather than owners because of the larger business they can conduct on their limited capital." To require landownership in order to farm would effectively seal off the opportunity to purchase land from families who did not start out with sufficient capital to invest in land before starting to farm.[30]

The issue of the relationship between tenancy and capital was debated by the American Association for Agricultural Legislation, a think tank and lobbying group based at the University of Wisconsin. They published arguments between Spillman and others in 1919, in which he pointed out that tenants wanted ownership and were willing to sacrifice their income and standard of living in order to own their own property. He believed the desire to own a home, the economic independence, and, ironically, the rising values of property made it an attractive investment. He considered it "fortunate" that tenants wanted to become owners as quickly as possible, because "a farming community made up entirely of tenants seldom maintains itself at as high a level of rural welfare as is the case with a similar community made up of farm owners."[31]

30. Spillman, Charles L. Stewart, and B. H. Hibbard, "Land Tenure and Public Policy—Discussion" (1919), 230; Benedict, *Farm Policies*, 116.
31. Spillman, Stewart, and Hibbard, "Tenure and Policy," 77.

Nevertheless, farm tenancy had its proponents who saw longer tenure as tenants as a positive reflection of the landowner's fairness. After all, if tenants were unhappy, wouldn't they leave the farm quickly? An example was the owner who had kept five families on his three farms for sixteen years, which was lauded as an example of fairness and stability. Tenants who moved from one farm situation to another within five years were viewed as "chronic shifters," and "shifting" of tenants was discouraged. The community had a responsibility to "exercise social control over its shifting, so as to cut down the cases of preventable shifting" in order to maintain social stability.[32]

If the government were to reform landownership, a standardized system of land policy appeared anathema to both Spillman and his critics. Reforms such as requiring all farmers to be landowners and landowners to work their own lands would eliminate the agricultural ladder and thereby take away all opportunity for young families to work their way toward ownership without capital. The tenancy debate continued into the Depression over whether share tenancy was a rung on the agricultural ladder or "whether the ladder is climbable and the rung is desirable," noted Joseph D. Reid Jr. at the University of Chicago. Studies fell into the two schools of thought; governmental policy makers adhered to Spillman's views (so did the Grange), and capital interests viewed share tenancy as a modern way to coordinate labor and capital. Studies done during the Franklin Roosevelt administration found share tenancy to be inefficient, wasteful, and evidence of failed land and agricultural policies.[33]

Spillman and Rockefeller Again

Once he left the Department of Agriculture, Spillman freely criticized federal policies in speeches and through the agricultural press, causing more controversy. In 1919, he delivered a speech critical of Secretary of Agriculture David Houston, which was disseminated through farmer groups across the country. One group, the Farmers' Union of

32. Ibid., 68.
33. Ibid., 76; Reid, review of *Agriculture in the U.S.A: A Documentary History*, ed. Wayne D. Rasmussen, 554.

Whitman County, in eastern Washington State, gathered to read and discuss the speech and resolved that the "press reports seem true and reliable" and that Houston had concealed and falsified reports relating to the production of wheat. The farmers noted that there "seems to be a higher power than the U.S. government over the Secretary of Agriculture, in the person of the Rockefeller General Education Board and also over many of the educational institutions of the Northwest with sinister motives, to mold public opinion for the benefit of the few at the expense of the many, which is autocracy and hypocrisy, against true and loyal democracy." The group requested an investigation of the department by Congress "to determine the facts, to deal with them as good, loyal government officials should with vice, disloyalty and malfeasance in office, and to see that the Rockefeller General Educational Board or any other combination keep their hands off the governance of food product prices." They dispatched the resolution to their members of Congress.[34]

Although little notice would accrue to such activities in an isolated group of disgruntled farmers, the allegations Spillman hurled at Houston gained attention from not only the farm community but industry and the department as well. Within a month of the Whitman County farmers' meeting, articles appeared in the national press, claiming Houston was a Rockefeller subordinate. Alfred McCann, a popular investigative journalist in New York, wrote at least two stories about the allegations. He noted that the "startling charges" made by Spillman had been discussed in farmers' meetings across the country and "will probably lead to a general housecleaning in Secretary Houston's department."[35]

The issue that Spillman publicized involved Houston's lying to the National Livestock Association, telling them the department was busy working on an investigation of the cost of producing farm products and that he was pushing it vigorously. Spillman revealed, however, that Houston had months earlier ordered the department to suppress any

34. "Whitman County Farmers Union: Resolution Passed Demanding That Congress Investigate the Acts of Secretary of Agriculture," *Pullman Herald*, February 28, 1919.

35. McCann, "Secretary Houston Attacked"; "Farmers' Board Favors Agriculture Dept. Shake-Up," *New York Globe and Commercial Advertiser*, March 29, 1919. The Rockefeller Foundation responded quickly, announcing two days later a foundation to provide half a million dollars for scientific research at 130 universities across the nation ("Rockefeller Science Fund," *New York Globe and Commercial Advertiser*, March 31, 1919).

work on cost-of-production figures. Houston had tried "to put a stop to" efforts of the Federal Trade Commission to "bring order out of chaos between cattle producers and the packers." That was not as worrisome, McCann noted, as the assertions that Houston was a tool of the Rockefeller General Education Board. Spillman told farmers that, when Houston took over, he had circulated a statement "representing Mr. Rockefeller's views, in which Secretary Houston concurred." Those views were adamantly against any investigations to determine the cost of producing farm products. Spillman claimed that Houston had said, "No representative of the Department of Agriculture should ever, under any circumstances, even intimate that it is possible to overproduce any farm product." Spillman revealed his frustration at the many bulletins the department refused to publish: "Anyone connected with the division of publications can testify to the difficulties encountered by manuscripts that relate to any phase of farm profits or farm costs."[36]

Spillman claimed Houston's link to the Southern Education Board, a subsidiary of the General Education Board, before he became secretary was significant. Spillman described the GEB's past involvement in funding the department, which had ended with the congressional investigations in 1914. Spillman reported that Houston had pledged to destroy the Office of Farm Management when he took office and that he had reduced funding by one-third, establishing a bureau inside the department termed the Rural Organization Service, which Spillman claimed was established by "the Rockefeller people." The work of the Bureau of Markets was to be placed under the Rockefeller bureau.[37]

Spillman explained that he "had nothing personal to gain" by exposing Houston, and "possibly I have much to lose." He believed strongly that "the public welfare is at stake." Spillman's charges, "because of his prominence, cannot be disregarded," the Dairymen's League noted. The National Board of Farm Organizations, made up of agricultural interests in thirty states, believed the charges corroborated what they already believed to be true. Congressman Louis Crampton of Michigan announced he would begin a congressional investigation. McCann supported Spillman's claims, too, noting he had talked with scientists in the department "who literally groan under the stifling restrictions heaped

36. McCann, "Secretary Houston Attacked."
37. Ibid.

upon them in an effort to prevent their findings from stirring up the stockholders of food establishments." Other bureaus in the department had been compromised, too, he noted, pointing to the Bureau of Chemistry and the Bureau of Animal Industry in particular. The "denaturing of official manuscripts" in the department had "become a public scandal," McCann claimed.[38]

The department did not take the charges lying down. Before McCann's articles appeared in New York, Assistant Secretary of Agriculture Clarence Ousley—"Colonel Ousley"—distributed an eleven-page letter to members of the press as well as to numerous employees across the country who worked under the direction of the Bureau of Markets. In the letter, Ousley attacked Spillman, claiming, "Mr. Spillman seems to have lost the capacity to distinguish between truth and error. Some of his statements are gross exaggerations, some are insinuations without evidence, and others appear to be downright inventions. It almost seems that Mr. Spillman has the mental habit of some children who imagine dramatic episodes and then describe them until they believe them to be true."[39]

Ousley pointed out he had known Houston for fifteen years or more. "I was associated with him in educational undertakings in Texas," he noted (through the General Education Board, he failed to add). Ousley explained he was under no obligation to Houston, as he had recently submitted his resignation to return to Texas anyway. He said Houston had been "over-indulgent with Mr. Spillman's frailties and shortcomings, and that Mr. Spillman, after the manner of many ambitious and disappointed men who cannot have their way, has resorted to unworthy methods of revenge." Ousley defended Houston, claiming he had never objected to cost-of-production studies but that he had objected to "inadequate studies and misleading data." Ousley admitted that "it is true that studies for the Federal Trade Commission were held up for a time; but this was due solely to a misunderstanding of what the Secretary had in mind." Ousley called Spillman a "demagogue" for claiming

38. McCann, "Farmers' Board Favors Shake-Up."
39. Letters and memos in Office of Bureau of Agricultural Economics, Correspondence Files, Records Group 83, USDA Archives. Ousley's letter and the "Memo of Understanding" are found in box 205, file 613-A, "A Memo of Livestock" PI 104 E, 3 170/32/17/1, General Correspondence, 1912–1922, Records Group 83, Records of the Bureau of Agricultural Economics, ibid.

that John D. Rockefeller had any influence in the department and said Houston had been too easy on Spillman. Houston had been aware of Spillman's "incompetency as an administrator and investigator," Ousley stated, and "it became clear that he was at least lacking in judgment. . . . [H]e was a theorist without sound judgment . . . a man of great vanity." Ousley said that Houston had met Spillman only once before he came to Washington (and, indeed, Houston had met President Wilson only once before joining his cabinet, too) and that he had no preconceived plan to destroy the Office of Farm Management. Ousley insinuated that Spillman was jealous, without mentioning that Spillman had been suggested by the press as a major contender for Wilson's cabinet, whereas Houston was a complete surprise.[40]

The "reorganization" Houston had done to the Office of Farm Management had actually put it (and Spillman) directly under Houston's control. Spillman became the only division head who did not manage a bureau of his own. Houston also took the farm demonstration work from under Spillman's direction and placed it under the States Relation Service, a unit Houston created. Houston also held down clerks' pay in the Office of Farm Management, keeping it at nine hundred dollars while elsewhere clerks earned twelve hundred dollars annually, which created low morale and constant transfers out of Spillman's office. Houston claimed the staff turnover in the OFM was because the employees did not have enough work to do, another arrow hurled in Spillman's direction. Ousley attached a copy of the 1906 "Memorandum of Understanding" between James Wilson and the General Education Board, which revealed nothing but did substantiate Houston's claims that the department was already involved with Rockefeller before he came aboard in 1913.[41]

Ousley also made public a letter from a wheat researcher, J. H. Arnold, who claimed the wheat studies Arnold had performed for cost-of-production data while working in Spillman's office were insufficient, based on personal observation, and based on "very meager statistical data." Arnold's letter was used to bolster support against Spillman because Arnold had worked directly under him, and a copy of the letter was included in the materials Ousley sent out to the press. Arnold

40. Ibid.
41. Ibid.

later felt great regret about what he believed was "selling out" Spillman, whom he respected. His suicide was later attributed to despondency over the issue.[42]

In April, the issue resurfaced in rural newspapers. The *Pullman (WA) Herald* reprinted the entire eleven pages Ousley had sent out without comment. Appearing next to the lengthy article was a quarter-page advertisement for Standard Oil gasoline.[43]

Farmers had been badgering Houston for cost-of-production information, both during and after World War I, but he resisted, saying, "The farmer is not entitled to any information on the cost of production. His business is to produce." Farm groups such as the Grange, Alliances, Farmers' Unions, and the American Society of Equity had proposed limiting production by estimating acreage needs and adjusting planting accordingly as early as the 1890s. As secretary of agriculture, David Houston saw no need for governmental involvement in adjusting agriculture, remaining optimistic about its future prosperity. Houston, however, "was viewed with suspicion, if not outright hostility by the farmers," historian Theodore Saloutos explains.[44]

By 1920, David Houston's career in agriculture was over. His attempts to remake the department failed, and in February President Woodrow Wilson appointed him to a different cabinet position, secretary of the treasury, and later head of the Federal Reserve Bank—a position he had always wanted and a move carefully made in the months before Wilson's term ended. Wilson appointed Edwin Meredith, publisher of *Successful Farming*, to fill the agricultural secretary position for the months remaining in Wilson's administration.[45]

The *Country Gentleman* series concluded with suggestions for improving the department. William Harper Dean noted that administrative positions were paid the highest salary, which attracted the best scientists, but "administrative jobs kill scientists." The research positions

42. Arnold to Houston, October 26, 1918, copied in Ousley to E. E. Miller, March 11, 1919, in ibid. It also appeared in newspaper stories that carried the full press release. See R. Spillman, "Biography of Spillman," MASC, 323. Arnold's suicide is not to be confused with that of Beverly Galloway, talented head of the Bureau of Plant Industry, who committed suicide in 1938. The two incidents are unrelated.

43. "Ousley Answers Spillman Charges," *Pullman Herald,* April 11, 1919.

44. Saloutos, *The American Farmer and the New Deal*, 15, 35, 17.

45. Houston, *Eight Years with Wilson's Cabinet*, 61.

needed better salaries, greater freedom, and fewer assistants (and administrative tasks).

> We must stop measuring the ability of a man by the genius he displays in securing large appropriations, the greater part of which must be spent in salaries for clerks and assistants and for his time in keeping them busy.... Before the Department of Agriculture can hope to cooperate successfully with the states it must look first to the adjustment of its own machinery at Washington. Before it can hope to cooperate with any individual or organization it must learn to cooperate with itself. For it is becoming an unwieldy machine.[46]

That situation changed under Henry C. Wallace, who brought life back to the department, in the words of Russell Lord, who compared Wallace's administration with that of James Wilson, politely ignoring the Houston years. Lord wrote:

> He went even further than [Wilson] in assigning able men to important chiefships or special assignments, sparking them with ideas to supplement their own, then crediting the whole result, if it came out right, to them; or backing them up, silently and steadily, if it didn't. Men who have been in the Department a long time observed that under most secretaries they often felt that they were ghosting for a passing figurehead as Secretary; whereas when you worked under H. C. Wallace he was up there in the front office, from eight in the morning until six in the evening, coolly taking the heat of blasts from the White House, the Capitol, and elsewhere, absorbing the blame with neither haste nor worry, passing the credit to subordinates—ghosting for them.[47]

In 1921, Spillman quickly returned to the department, when Henry C. Wallace invited him back to act as a "consulting specialist in farm management," a position that gave him the independence he sought. He was invited back to the department "at the maximum salary the law allowed, with a commission as a free lance to do whatever interested

46. Dean, "What's the Matter with the Department? The Confusion of Tongues," 590.
47. Lord, *The Wallaces of Iowa,* 224.

him," his son, Ramsay, noted. Spillman was free from executive duties and totally unencumbered by restrictions. For a decade, Spillman wrote, researched, traveled, and lectured. He spoke about agriculture on the Chautauqua circuit, wrote a high school agricultural textbook, developed the law of diminishing returns and published a book about it, and developed his ideas about domestic allotment, publishing a book about it as well. He also taught classes in geography, international trade, and agricultural policy at Georgetown University's School of Foreign Service, where he was on the graduate faculty. Secretary Wallace's direction allowed him to do research on a national, even international, scale while continuing to shape policy and higher education.[48]

48. R. Spillman, "Biography of Spillman," MASC, 346.

4

The Law of Diminishing Returns

I n the early twentieth century, fertilizers, like genetics, shifted from traditional methods to scientific applications. Spillman, a scientific agriculturist, was again involved in the dynamic transformation of scientific principles to practical utility. Because he clearly described how natural laws could be used by farmers to make better decisions, he regarded his publications on fertilizer application as the most important work of his life. His book *The Law of Diminishing Returns* was essentially about fertilizer, but he advocated a significant departure from agricultural practices by challenging ideas regarding how much fertilizer could do to effect increased yields. In this case, he identified a mathematical model that is still valuable in economics, the law of diminishing returns.[1]

The accelerating use of artificial fertilizers after World War I challenged Spillman's long-held belief in the importance of maintaining humus in the soil, notably practices that protected and replenished the soil, such as crop rotations, planting clover as a cover crop and green manure, and spreading animal manure. The artificial fertilizer industry promised to change that dynamic: farmers could grow abundant crops on even poor soils if they applied chemical and mineral additives to the field. This meant farmers with more capital to spend could apply more fertilizers, reaping greater harvests. Diversified farming, which Spillman had proselytized for so long, appeared outdated, particularly when one advocated using science to foster agriculture. Rather than resist change,

1. R. Spillman, "Biography of Spillman," MASC, 297.

however, the challenge artificial fertilizers presented to Spillman's scientific thinking led him to determine how commercial fertilizers could be more successfully applied.

Fertilizers in Perspective

Fertilizer history begins about two hundred years ago with the recognition that plants utilize nutrients from the soil. Concurrent with Thomas Malthus's prediction that the world's population promised to outstrip its resources, chemists identified elements that plants absorbed during growth, launching the plant-food industry and expanding agricultural production. In 1840, German chemist Justus von Liebig was the first to experiment by adding plant nutrients to soil. He believed that crop yields increased or decreased in direct relationship to additions or reductions of mineral nutrients. His theory was widely accepted, and he was the first to advocate the application of fertilizers.

Liebig turned soil science in a new direction when he recognized that plants needed not only organic components of soil (the "humus theory") but also inorganic substances such as minerals, as well as carbon from the air. While continuing work on soil chemistry, Liebig realized that adding nutrients to soil did not increase yields in direct proportion. Calling it "the law of the soil," Liebig recognized that additional inputs failed to produce proportional effects, an idea E. A. Mitscherlich and William Spillman would later build on. Liebig recognized what he called the law of the minimum, in which yield depends on the factor that is at a minimum, or the necessary input in least supply. For example, water might be the minimum factor in nonirrigated crops. No matter how many soil components or how much sunshine is available, crop yield in this case will be limited by the amount of water available.[2]

Nitrogen, potassium, and phosphorus were the first and most important crop nutrients to be recognized. Nitrogen was found in nitrate deposits in Chile, potassium could be obtained from potash, and phosphorus came from ground-up bones or phosphate rock.

Artificial fertilizers actually grew out of the manufacturing industry. The by-product or waste after steel production visibly boosted plant

2. Spillman and Emil Lang, *The Law of Diminishing Returns* (1924), 136.

growth when spread on soils, so fertilizer production became an adjunct of the steel industry. Treating bones or mineral phosphate with sulfuric acid (another manufacturing by-product) created a valuable crop fertilizer, called superphosphate. By the mid-nineteenth century, superphosphate was widely manufactured in Britain and the United States from bones. As bones became less available, mineral phosphate was substituted for bones to create superphosphate. Superphosphate, made from industrial by-products, supplied phosphorus, an essential element for plant growth, but plants also need nitrogen and potassium.[3]

By 1900, American farmers began importing nitrogen fertilizers from Chile, where mineral sodium nitrate was abundant along the coast. The valuable nitrate (sometimes called saltpeter) deposits, initially found on Bolivian territory, incited one of Latin America's fiercest wars, as Chile fought Peru and Bolivia over the region, which was finally ceded to Chile in 1881. Fertilizer exports became the backbone of the Chilean economy, with a tax on nitrate exports making up 80 percent of Chile's total revenue.[4]

Potassium, the third essential plant nutrient, came from potash, a mineral salt deposit found (at that time) only in Germany. Potash strata ran twelve hundred to thirty-five hundred feet belowground, in layers five thousand feet thick, left by evaporation of ancient saltwater lakes. Before World War I, Germany supplied the world demand for potash, exporting more than eleven million tons annually.[5]

A New Fertilizer Theory

Scientists long accepted Liebig's idea that to improve soil's fertility, certain components of the soil had to be balanced. A certain proportion of nitrogen, phosphorus, and potassium could be identified and adjusted so that the soil itself was in appropriate balance. Because the soil was viewed as key to producing better plants, chemists focused on soil composition but ignored plant needs. In 1919, a series of articles

3. T. K. Derry and Trevor I. Williams, *A Short History of Technology: From the Earliest Times to A.D. 1900*, 553.

4. Ibid., 553.

5. *The Potash Industry*, 34.

about artificial fertilizers appeared in the *Farm Journal* with an introduction by Spillman. The articles took a new stance on fertilizers, asking farmers, "What are we raising, anyhow—soil or plants?" Plants were the focus, and the author, A. B. Ross, pointed out that plant needs should be used as the basis for choosing soil amendments.[6]

Ross, a former attorney who had taken up rural living in Pennsylvania on the advice of his physician, was fascinated by geology. He realized that his farming neighbors were getting meager returns for their work and that the farmers he knew were stuck in poverty. He discovered that many farmers would have nothing to do with the state experiment station, "which they considered as an institution having unlimited resources with which to experiment, and one carrying on farming under conditions which the ordinary farmer, for lack of money, could not hope to imitate; and they did not read either state or national bulletins because they could not understand them." In 1910, Ross took it upon himself to act as a volunteer interpreter of the bulletins and engaged several farmers in working experiments on their own places in Bedford County, Pennsylvania. Eventually, he was working with nine hundred farmers, at his own expense. He needed financial assistance or he would have to return to practicing law, so he contacted the Department of Agriculture, where Spillman hired him to work for the Office of Farm Management. Ross had pretty much invented the "county agent" idea on his own, before coming to the department. So, when Spillman, as editor, gave his glowing recommendation to the *Farm Journal*'s series written by Ross, he was lauding one of his former colleagues, whom he had hired and trained nearly ten years earlier.[7]

The new theory looked at what nutrients the crop (not the soil) needed, then added those to the soil. Claiming the "effort to balance the soil is bare, bald, unfruitful, academic theory," Ross argued that scientists were "asking the question of the soil which it can not answer." Asking the question of the plants, however, made things more successful. Ross noted that once the right amounts of essential nutrients were identified for each particular crop, commercial fertilizers could be applied successfully. "It means that with commercial chemicals, we can

6. Ross, "Old Fertilizer Theories All Scrapped," 10.
7. R. Spillman, "Biography of Spillman," MASC, 39.

farm without manure and at a cost well within the limits of profitable farming, if these chemicals are used alright." The use of animal manure was declining as farmers moved to power farming and away from draft animals.[8]

Ross's statements were highly controversial. He argued that balancing chemicals to match plant needs allowed farmers to use *less* of the commercial fertilizers they were applying. Studies revealed that in some cases farmers were applying far too much of certain components, which actually diminished their yields. For example, farmers had been applying nitrogen to plants such as peas and beans that had no need for it; the bacterial action in the soil that produced nitrogen in legume roots was hindered when commercial chemical nitrogen was applied to the field. Ross argued that farmers who added too much phosphoric acid and nitrogen actually diminished their yield, on top of paying additional costs for the commercial fertilizers. "We have been confusing extra production with extra profits," one article in the series pointed out.[9]

The fertilizer series ended in May 1920, with advice to "prove the new fertilizer facts on your own." Indeed, the information was what farmers had been clamoring for. The judgment about fertilizer application had been much like that about other inputs and expected returns: the more one invested, the greater the increase in output. But farmers had realized something was amiss; sometimes fertilizer application did not pay for itself, and other times the outcome was unpredictable. The *Farm Journal* series broke new ground, so to speak, pointing to the need to understand plants as well as soil and the realization that a one-size-fits-all theory for fertilizer application was not optimal. Farmers had been told by fertilizer salesmen, and perhaps based on their traditional use of organic fertilizers, that the more amendments added to the soil, the better it became. That worked with manures and plant-based amendments such as composts, but chemical additives, such as sulfate and nitrogen, did not give the same results. In fact, as the series of articles pointed out, in some cases adding too much of a commercial fertilizer actually decreased yields.

8. Ross, "Old Fertilizer Theories," 10; Ross, "Commercial Nitrogen—the Great Gold Brick," 12.
9. Ross, "Commercial Nitrogen," 8.

Fertilizers and War

One reason farmers were using so much commercial fertilizer was because the supply, so long restricted by cost, transportation, and foreign
governments, had suddenly increased after World War I. In 1913, German chemists figured out how to fix nitrogen from the air, through a
synthetic ammonia process using coal to synthesize nitrogen, which
freed them from reliance on Chilean nitrate sources. But the process
was secret, and U.S. scientists were stymied. In 1914, a laboratory at the
Department of Agriculture's Arlington Farm opened, where research
was eventually subsidized by the War Department. "It is ironic, perhaps,
that the first great expansion of nitrogen production was carried out to
meet not the needs of agriculture, but the demands of war," notes economist Mirko Lamer.[10]

In 1916, Congress discussed the nitrogen supply because it was both an
agricultural and a military problem. Nitrogen, essential in manufacturing gunpowder and explosives, was essential to national security. It was
clear that future warfare would depend upon the supply of nitric acid for
explosives that could be created by manufacturing synthetic nitrogen.
The Chilean sodium nitrate deposits might not last indefinitely—the
supply was exhaustible, and the United States alone paid Chile sixteen
million dollars in 1915 for nitrates. The Panama Canal, a boon for easier
access to the Chilean coast, might be closed during wartime, making it
even harder to import nitrates from the west coast of South America to
southern U.S. cotton farmers, the major American users of nitrates.[11]

By 1919, the War Department established the Fixed Nitrogen Laboratory at American University in Washington, D.C., turning it over to
the Department of Agriculture in 1921. By the midtwenties, an American
air-nitrogen industry based on the direct synthetic ammonia process
was well established, and farmers were able to reap the benefits of what
was initially a wartime effort.[12]

War also stimulated a search for new supplies of potash and phosphate to protect national food supplies. The importance of national self-

10. Lamer, *The World Fertilizer Economy*, 4.
11. *Congressional Record*, 64th Cong., 1st sess. (April 14, 1916): 6144–46.
12. T. Swann Harding, *Two Blades of Grass: A History of Scientific Development in
the United States Department of Agriculture*, 204.

sufficiency led to discovering additional mineral reserves, improved fertilizer manufacturing methods, and increased production and use of fertilizers. A plant at Muscle Shoals, Alabama, was the initial center for processing phosphoric acid from low-grade phosphate ores, using an electric furnace process. The first processing plant of this type opened at Muscle Shoals in 1920, operated by USDA-trained staff.[13]

At first, U.S. scientists created nitrogen plants using large amounts of hydropower energy, but after World War I ended they had access to Germany's cheaper synthetic ammonia method. After World War I, Germany's Alsace-Lorraine potash deposits were given to France; at the same time, potash deposits were developed in the Urals, Palestine, and Poland. In the United States, deposits in Searles Lake, California; Utah's salt flats; Nebraska's salt lakes; and Montana provided domestic alternatives.[14]

After the war, the emphasis on fertilizers for explosives as well as agricultural production facilitated discoveries of new deposits as well as new manufacturing techniques. Chilean nitrates dropped in price because ample amounts of ammonium sulfate and calcium cyanamide (from Canada) created competition. Prices for Chilean nitrates had soared during the war, going up more than 230 percent between 1910 and 1918. When wartime demands for nitric acid for explosives declined, so did prices, and by that time domestic production was well under way. In the aftermath of World War I, fertilizer supplies were substantial, prices were competitive, and the industry moved to entice farmers to use as much as they could afford.

Farmers and Fertilizer Use

"Fertilizer practice[s] are determined more by the stage of agricultural development and commercial propaganda than by the intrinsic needs of soils and crops," writer E. M. Crowther observes. Getting farmers to use fertilizers did not depend upon their understanding of

13. Ibid., 205.
14. Lamer, *The World Fertilizer Economy,* 109; "Consumption of Fertilizers in the Southern States"; Charles H. MacDowell, "Problems and Processes in Mixed Fertilizers," 72.

plant or soil needs; rather, fertilizer use corresponded to their economic situation and to trends in the agricultural economy. Although fertilizers appeared to fulfill the promise of scientific agriculture, farmers' use and understanding of fertilizers were seldom based on science.[15]

Convincing farmers to adopt commercial fertilizers was not difficult; increased yields spoke for themselves. Although extension agents offered advice about fertilizers, most farmers followed advice from fertilizer salesmen; more than one-half of the fertilizers used in the United States were chosen due to salesmen's recommendations. Indeed, salesmen's advice was followed even when it contradicted findings exhorted by experiment stations. Farmers applied potash and phosphates at high rates suggested by eager marketers, ignoring the more conservative recommendations from most extension agents.[16]

Along with eager salesmen extolling fertilizer's virtues, cotton growers found it was a weapon against insect pests. It was the boll weevil that pushed southern farmers to adopt commercial fertilizers. Manufacturers skillfully used southern extension agents to spread their message that fertilizers would save cotton crops from the boll weevil. "The staff has continued to center its activities in no small measure against the boll weevil," a trade publication for the Southern Fertilizer Association noted. "Nothing brings out farmers like an announcement of a boll weevil meeting. . . . [T]he subject is broad and opportunity is presented to give fertilizers emphasis as an important factor in meeting the insect. . . . [N]o better opportunity could be found for promoting the intelligent use of fertilizers than is afforded by these boll weevil meetings." The methods for fighting the boll weevil had roots in Seaman Knapp's demonstration farming techniques, which included liberal use of commercial fertilizers. Nitrogen added early in the plant's life cycle promoted quick growth, phosphoric acid pushed the plant to form bolls quicker, and potash strengthened the plant, helping it resist weevil damage. Amazingly, fertilizer applications allowed a cotton farmer higher yields, even when weevils infested the field, because the chemicals sped up the plant's growth cycle before the weevil could destroy it. By 1917,

15. Crowther, *Fertilizers during the War and After*, cited in Lamer, *The World Fertilizer Economy*, 31.

16. Lamer, *The World Fertilizer Economy*, 125.

the fertilizer strategy against the weevil was no longer necessary; calcium arsenate poison eliminated the bugs. But fertilizers had secured a place in southern agriculture, boosting production as if by magic.[17]

The "fertilizer controversy" following the *Farm Journal* series went unnoticed by readers, but W. J. Spillman found himself in hot water. As editor of the series, he had written a prominent sidebar beside the first article, with the attention-getting title "Old Fertilizer Theories All Scrapped." Spillman's introduction stated that the articles would "revolutionize the use of commercial fertilizers," adding that the information would "spell the doom of commercial nitrogen in rotation farming, triple or quadruple the demand for potash, and stand the whole fertilizer situation on its head." Spillman's words were strong: "Millions and millions of dollars have been wasted by farmers in following the principles of fertilization laid down by Liebig in 1842." Indeed, farmers had been hurting their cause with excessive use of phosphorus and nitrogen, actually diminishing crop returns. Spillman emphasized: "We regard this article of Ross's as the most important thing we have ever seen on the subject of commercial fertilizers, and *The Farm Journal* may well be proud of having printed it first." He was astute about the implications for industry, predicting that "we will no doubt be jumped on hard for this." He knew commercial interests would not like the concise information about fertilizers and their effects (or lack of them). "It throws a firebrand into the camp of the chemist, and plays hob with the phosphate people as well as with the nitrate people. It also plays into the hands of the potash people," he added. He realized it would hurt some business interests, while wildly assisting the potash industry. And potash was largely an imported product. He noted that because potash was so important, "it makes it more important than ever that we get from under the German potash monopoly."[18]

Spillman was right; criticism was not long in coming. Several abusive letters arrived, including some from Cornell faculty with whom he had once worked closely. That summer Cornell sponsored a fertilizer course for 110 fertilizer industry executives, which the industry hoped

17. Southern Fertilizer Association, *Yearbook* (1924), 16; *American Fertilizer* (February 14, 1920), 122.
18. Ross, "Old Fertilizer Theories," 10.

would influence the agricultural colleges to expand fertilizer consumption. The National Association of Fertilizer Manufacturers (NAFM) was not shy about its interests. A serious advertiser in the agricultural press, the NAFM sent the chairman of an affiliate, the Soil Improvement Committee, and three or four other executives to the *Farm Journal* office to demand retraction of Spillman's piece. The magazine held its ground, and Spillman's only concession was a less inflammatory headline on the next installment: "Commercial Nitrogen—the Great Gold Brick: *In Rotations Containing Clover.*" Clover was a "green manure" plant, which provided nitrogen to the soil through root nodules. Indeed, clover needed no application of commercial nitrogen to enhance growth.[19]

Interestingly, the fertilizer series created a surge in circulation, moving the magazine to two hundred thousand readers. The fertilizer people increased their advertising, going from an occasional one-third page to a full page. Word got back to the editors that "the *Farm Journal* advertising during this controversy was the most productive advertising they had." During the period between 1903 and 1913, fertilizer usage had increased by 7.5 percent per year; between 1921 and 1927, annual fertilizer consumption increased by more than 11 percent. The types of fertilizer changed, however, as farmers purchased higher-grade products, as well as increased amounts of potash. In the three years following the *Farm Journal* fertilizer series, sales of rock phosphate in Illinois had dropped from one hundred thousand tons per year to two to three thousand tons annually. Farmers realized it was a waste of money to apply it to nitrogen-fixing crops such as alfalfa and legumes. Although they became more discerning about fertilizers, they continued to adopt fertilizers, even as they cut back on acreage, finding it a solution to low prices and overproduction. Lowering the cost of production with intelligent use of fertilizers became crucial to developing twentieth-century agriculture.[20]

But the *Farm Journal* series was only one voice in the cacophony surrounding the new era of commercial fertilizers. The fertilizer industry's trade group, the Southern Fertilizer Association, headquartered in

19. *American Fertilizer* (February 14, 1920), 61.

20. *Nitrogen,* 38; Southern Fertilizer Association, *Yearbook* (1924), 19; Spillman Papers, cage 250, box 3, file 23, MASC; R. Spillman, "Biography of Spillman," MASC, 341.

Atlanta, funded the Soil Improvement Committee as its promotional arm. The Soil Improvement Committee developed a systematic, wide-reaching, and effective program to overcome any resistance to commercial fertilizers. Their major foe was the agricultural extension and agricultural college community, which promoted legume crop rotations, manure, and small amounts of fertilizers, rather than the huge amounts industry recommended. The Soil Improvement Committee mailed large numbers of charts and brochures and made sure industry speakers were included at all agricultural college events. They provided articles for the farm press, honored county agents who promoted fertilizers with trips to the American Society of Agronomy meeting, sent press releases to five hundred farm newspapers, and sponsored continued pressure on agricultural teachers to promote the use of fertilizers.[21]

The Law of Diminishing Returns

Once fertilizers were accepted, questions arose. How much should one use? Was more always better? Could one predict how much increased production to expect for a given amount of fertilizer? Farmers had not always been pleased with results from fertilizer applications, but no one knew whether they were using too much or too little. "Vast sums of money have been expended for fertilizers in the South during the past forty years," *American Fertilizer* noted. "The results have not always proved satisfactory." The problem was not with the fertilizer, but with the lackluster amounts farmers had applied, according to the industry. "The controlling element, in a large majority of the cases, was the uneconomical practice of applying too small a quantity of fertilizer.... [S]ixty-six to seventy of every 100 farms [in Georgia] are operated by tenants and they uniformly use fertilizer sparingly." One official at the Georgia Department of Agriculture suggested farmers apply one thousand pounds per acre on cotton land. In the past, farmers buoyed by high crop prices due to export demands during World War I had limited fertilizer applications only by their ability to pay for them. Spillman, through the *Farm Journal*, admonished that "the quantity of manure we have at our disposal is limited; the natural tendency is to

21. Southern Fertilizer Association, *Yearbook* (1925), 18.

make it go as far as possible. But the quantity of commercial fertilizer at our disposal is limited only by our willingness to buy it. Our willingness to buy it must be just as big as the opportunity for profits in its use requires—if we are farming for profit." Clearly, someone had to figure out how much was successful and profitable.[22]

Because Germany had long been the center of both chemical and fertilizer science, it is no surprise that German chemists were the first to develop synthetic fertilizers from coal and also the first to try to figure out how crop yields related to increasing applications of fertilizer. E. A. Mitscherlich studied yields of rye, a staple crop in Germany, and found that with increasing applications of fertilizer, the yield increased at a rate that grew smaller. When one identified the amount of increased yield for each increasing amount of fertilizer, it was easy to see that the more fertilizer one applied, the less the yield increased. Certainly, the yield increased, but not in proportion to the increasing amount of fertilizer. Mitscherlich created "Yield Tables" showing his results and set off a flurry of work by soil scientists trying to either prove or disprove his work.[23]

Mitscherlich's formula was criticized because it did not consider other variables, such as climate, crop varieties, or soil condition. The idea that a law of diminishing returns might portend nonsustainable agriculture seemed so far-fetched that it had many critics. It was officially denied in the Soviet Union where Marxist theory held that agricultural productivity could be increased with human creativity and effort. The Soviets had only begun adding fertilizer and focused heavy attention on plowing methods, crop rotations, mechanization, and irrigation. The idea—that applying science to effect nature had limits—seemed anathema.[24]

Spillman had always advocated the scientific management of the natural world, and to him optimal fertilization involved crop rotations including nitrogenous plants such as legumes or clover, which would naturally inoculate the soil with nitrogen. Though Spillman's recommendations did not mesh with the expanding industrialization of agriculture, he did not oppose synthetic fertilizer use. Instead, he turned

22. *American Fertilizer* (January 31, 1920), 142; "Getting Down to Brass Tacks," 8.
23. Lamer, *The World Fertilizer Economy*, 73.
24. Ibid., 80.

his attention to figuring out a mathematical model for the law of diminishing returns in order to determine optimal fertilizer application, which he believed important because, in his judgment, farmers were being encouraged to use far too much commercial fertilizer. In December 1920, while at *Farm Journal,* Spillman sent an article, "A Plan for the Conduct of Fertilizer Experiments," to the *Journal of the American Society of Agronomy.* He argued that fertilizer experiments needed to look at multiple inputs, rather than just a single variable. He realized that fertilizer combinations were sometimes more effective than single elements alone. He explained, "The behavior of a fertilizer ingredient depends on the relative amounts of other fertilizer elements available to the growing crops."[25]

In the 1920s, Mitscherlich began tinkering with Liebig's theories, moving away from the long-held law of the minimum. He created a mathematical formula to express the proportional fading effect of additional fertilizer inputs past a certain point. In Germany, it was termed *Mitscherlich's law* or the *law of physiological relations.* Mitscherlich did not replace Liebig's law of the minimum; rather, he expanded it to include the relationship between components as well as the relationship of increasing inputs. Liebig's work led only to discovering which nutrient was lacking and how increased application of that component would increase yield. It was Mitscherlich who saw beyond the minimum component to develop a relationship among other elements and relate them to increasing inputs.

Spillman, by translating Mitscherlich's work, as well as Emil Lang's interpretation of it, brought the German theory to an English-speaking audience. In doing so, Spillman discovered the mathematical formula that both explained how increasing inputs declined over time and provided a means for relating the theory to useful applications. Spillman understood Mitscherlich's idea and developed mathematical models for the natural law, using examples from stock-feeding experiments to identify the predictable point after which additional inputs had little effect.

In the preface to his 1924 book, *The Law of Diminishing Returns,* Spillman describes his argument thus: "The increments in yield corre-

25. Spillman, "Fertilizer Experiments" (1921), 304–10; Spillman, "Measuring Absorbed Phosphates and Nitrogen" (1930), 215–16.

sponding to successive equal increments in fertilizer applied to a crop tend to constitute the terms of decreasing geometric series; thus, if R represents the ratio of the second increment to the first, then the ratio of the third increment to the second, of the fourth to the third, and so on, is also R." Spillman referred to his own version as the law of diminishing increment. The idea applied to "natural phenomena, both biological and physical," Spillman explained. He confirmed his theory by experiments with fertilizers and irrigation water, but noted that it also applied to fattening livestock and growth rates of children, as well capital and profit. Spillman thought the concept was useful to economics, agronomy, animal husbandry, anthropology, physiology, irrigation engineering, and mathematics. Indeed, it has been widely useful.[26]

His theory was based on the economic concept that by increasing a particular unit input, at some point returns from that input begin to decrease as the input increases. "The fact that the application of a second unit quantity of fertilizer does not ordinarily produce as great an effect as the first unit is an illustration of this law," he explained. A gradual increase in profit per acre occurred up to a certain point in the application of fertilizer. That exact point, after which additional applications of fertilizer became unprofitable, was the focus of his mathematical model.[27]

Spillman initially realized there might be a natural relationship when he saw a chart published in the *Country Gentleman* in 1921. It was a chart of fertilizer usage on cotton at an experiment station, showing decreasing yields with each successive application of fertilizer. "It occurred to me that possibly the figures in the second column tend to constitute the terms of a decreasing geometric series," he wrote. "If the second figure is, say, 77 percent of the first, then the third figure tends to be 77 percent of the second, the fourth 77 percent of the third, etc."[28]

Mitscherlich had discovered the relationship between crop yield and the increase in a growth factor (water, fertilizer, sun) and developed the following mathematical expression: $Y = A(1-e-kx)$, in which Y is yield per acre, x is the quantity of the growth factor available, A is the maximum yield obtainable by the use of any amount of the factor, and k is a

26. Spillman and Lang, *Law of Diminishing Returns,* iii–iv.
27. Ibid., vii.
28. Ibid., 1.

constant. Working from Mitscherlich's foundation, Spillman discovered the equation could be more effective if the following changes were made: $Y = A(1 - Rx)$. He wrote that in his model, "Y, A and x have the same significance as in Mitscherlich's formula, while R is the ratio of a decreasing geometric series, the terms of which are the increments of Y corresponding to successive equal increments in x." Y would be the amount of increased yield, A the sum of the series to infinity, R the ratio, and x the units of fertilizer (or whatever input was being measured).[29]

Spillman tested his theory by applying his algebraic equation to USDA statistics garnered from experiments feeding hogs, applying irrigation water, applying phosphate on cabbage fields, and feeding steers. His model promised to be invaluable in calculating exactly how much fertilizer to apply, replacing the industry's practice of applying as much as the farmer could afford to purchase.

But Spillman also saw it as a basis for further scientific work. "As is the case with any vital discovery in science," he wrote, "the discovery of the mathematical expression for the law of the diminishing increment is useful mainly in pointing out new fields of investigation—in suggesting new problems, and new ways of attacking old problems." Others tested his hypothesis. In 1931, scientists at the Bureau of Animal Industry applied his theory to animal feeding and found it both accurate and valuable. In an era when animal metabolism was not understood, Spillman's formula was the only scientific method for relating feed consumption and growth.[30]

Spillman's formula for figuring the law of diminishing returns has been widely applied in industry. From fish feeding to consumption studies and livestock feeding, the formula is reliable and well known. In 1977, Janusz Jaworski, of the Academy of Economics in Kraków, Poland, developed what has become known as the Spillman Production Function, by extending it and modifying its analytical and predictive properties, a new development on William Spillman's work. It is used in fishery economics as a bioeconomic model to predict outcomes. Jaworski wrote in the *Canadian Journal of Agricultural Economics* that he hoped

29. Spillman, "A New Basis for Fertilizer Experiments" (1930), 135–36.

30. Spillman and Lang, *Law of Diminishing Returns*, 33; Walter A. Hendricks, Morley A. Jull, and Harry W. Titus, "A Possible Physiological Interpretation of the Law of the Diminishing Increment."

his work on mathematical and statistical applications of the function would encourage others to research further on the classical Spillman model.[31]

Spillman not only accepted the use of chemical fertilizer amendments as part of practical farming but was also able to use scientific data to identify appropriate levels of use. The law of diminishing returns, a product of global scientific thinking and the application of scientific methods, was elegantly simple. He demonstrated that scientific reasoning was superior to haphazard guesswork and industry hype and ultimately benefited the public.

31. See Jaworski, "Decision Aspects of the Spillman Production Function."

5

Balancing the Farm Output

Farming had changed considerably between Spillman's arrival at the U.S. Department of Agriculture in 1902 and the aftermath of the Great War. Not only had science and technology changed agriculture, eliminating large numbers of draft animals, but the proportion of farmers who owned their land had diminished as well. Economic conditions in the postwar era affected farmers drastically, as their products found few buyers in a rebounding global marketplace rapidly embracing protectionism.[1]

Spillman had extolled better management of the soil to increase yields, which made sense in 1902. Better soil meant higher yields without expanding onto more land, which was getting difficult to do because of rising land values. Enlarging farm size depended on mechanization, but that necessitated increased capital investments in draft animals, and a lack of adequate seasonal labor made the former impractical for most farmers. Land was simply too expensive and difficult to obtain to allow for additional capital inputs for more men, mules, and machines. The "small farm well tilled" philosophy, which Spillman espoused in 1902, appeared sound until the nation experienced the export demands created by World War I and the surge in production that followed.

In August 1914, Secretary of Agriculture David Houston saw cotton production as crucial to maintaining the United States' balance of trade with Europe; yet with war erupting, how could exports continue unaffected? The United States owed three hundred million dollars in

---

1. Theodore Saloutos and John D. Hicks, *Agricultural Discontent in the Middle West, 1900–1939*, 373.

"floating indebtedness" to Europe, he pointed out, which was to come due in January 1915. His efforts pushed farmers to higher production levels very early in the war, particularly cotton growers, in order to balance the nation's trade deficit. In the fall of 1918, Houston urged them to increase wheat plantings for the next year. He chided them to "not let the war fail because of deficient food production" and to "show their fighting spirit and ability to meet serious situations" by increasing wheat plantings. "Let us fight in the furrows," he demanded. "The man behind the plow is the man behind the man behind the gun." Farmers had increased production in 1917 by 26 percent over the year before. Ten percent more acreage was in production in 1917, and the government urged more for 1918 and 1919.[2]

The *Banker-Farmer*, published by the American Bankers Association and directed at rural bankers, also extolled the importance of continuing to produce at high levels, even when the war was over. Clarence Ousley, assistant secretary of agriculture, urged the bankers that "We Must Not Stop!"—that the organized effort between banks and the Department of Agriculture, through county agents, must continue in order to "advance prosperity." He admitted, "Heretofore we have urged planting programs on the basis of patriotism. We can still appeal to farmers on the basis of humanitarianism, but we must now allow full freedom of choice from the stand-point of self-interest." How to keep farmers producing at high levels when they were not making a profit looked problematic. Nevertheless, "It looks to me like agricultural prosperity for a number of years, and I rejoice in the prospect," Ousley said.[3]

But as the war ended, the huge 1919 harvest found no export market. Dissent was growing as the government continued to urge farmers to full production. Herbert Hoover, head of the Food Administration, exhorted that "the war has been brought to an end in no small measure by starvation itself and it cannot be our business to maintain starvation

2. Houston to Albert Shaw, March 23, 1915, box 242, General Correspondence, Records Group 16, USDA Archives; "The Man Behind the Plow Is the Man Behind the Man Behind the Gun," U.S. Department of Agriculture poster, September 1919, box 620, ibid; Houston, "How the American Farmer Has Answered the Call of War and Work Ahead for Him Reviewed by Secretary Houston," *The Official Bulletin*, Committee on Public Information, Washington, DC (February 21, 1918), box 619, General Correspondence, Records Group 83, Records of the Bureau of Agricultural Economics, National Archives.

3. "The Outlook for Agriculture," 2.

after peace." The United States would be able to furnish about twenty million tons of food for export, compared to the six million tons prior to the war. He estimated that grain supplies had accumulated in Argentina, Australia, and other markets inaccessible during the war and that once shipping resumed those stores would hit the market, but he still recommended that "we have reasonable promise of ability through increased production and conservation to export seven times as much products as our pre-war average. . . . [W]e are justified today in our every act in the stimulation of production." Adding that "empty stomachs mean anarchy and disorder in Europe," he urged bankers to support farmers in continued full production. Insinuating that bankers might face problems at home, he explained, "From an ability to supply their people, grows stability of government and the defeat of anarchy. Did we put it on no higher plane than our interests in the protection of our institutions, we must bestir ourselves in solution of this problem." Protecting the stomachs of Europe would ultimately protect America's rural banks.[4]

Fear of bolshevism permeated postwar America. As Houston, Ousley, and Hoover admonished bankers to provide loans to prevent rural anarchy, even rural children were suspect. The "Junior Farmers" page of an agricultural magazine asked children these questions: "What would become of the world twenty years hence if the boys and girls of today all stopped studying and neglected to improve their minds and did not take advantage of opportunities for bettering their condition? What would become of the boys and girls? What would become of Progress? What would happen to Happiness and Prosperity? You know the answer. Don't let it happen. Don't be a *Bolshevik*."[5]

In the 1920s, U.S. agriculture suffered postwar blues. Between July and December 1920, corn prices fell 78 percent, wheat dropped 64 percent, and cotton was down by 57 percent. Livestock, too, fell from a high at war's end of $16.45 per hundredweight to $7.31 by the end of 1920. Surplus commodities found no market, land values plummeted, and bankruptcies soared. Net farm income fell from $9 billion in 1919 to $3.3 billion two years later. The situation was dire for many farmers who

4. Hoover, "A New World Food Situation," 4.
5. "Junior Farmers," 9.

could not pay debts or even plant the next year's crop. U.S. agriculture's golden age had ended; wartime expansion and rising commodity prices had come to a resounding halt.[6]

The war had pushed up production with a voracious export market, and, as historian Theodore Saloutos explains, farmers expected to maintain the high levels of production and profit they realized during the war and did not foresee how the situation would change in the 1920s. Overexpansion to fill wartime needs could not be readjusted quickly to the dwindling market, and the agricultural sector of the U.S. economy languished. Scientific advances in agriculture had bolstered production with better seeds and fewer losses to disease. Technology had increased mechanization and power, meaning more land could be cropped and less pastured, an increase of 15 million to 25 million acres more between 1918 and 1930.[7]

As the United States became a creditor nation for the first time and attempted to close off the domestic market to imports to protect prices, debtor nations could neither afford to buy U.S. commodities nor afford to repay wartime loans. Though the entire U.S. economy suffered postwar adjustment, the agricultural sector was especially hard hit, as exports declined due not only to tariffs but also to domestic recovery in Europe.[8]

Farming required more cash inputs as farmers became less self-sufficient. Neighbors could no longer rely on cashless practices such as sharing labor or animals to supplement their investments. Mechanization made cash imperative to purchase fuel, machinery, and parts. Farmers also needed more cash to pay taxes than ever before: taxes for rural improvements doubled between 1917 and 1930. By 1930–1931, one-fifth to one-third of farmers' net income went to pay taxes.[9]

Immigration restrictions and stagnant wages reduced domestic demand, and the rapid growth of urban sectors slowed. Farmers found themselves forced to increase production in order to cover their costs, creating an ever expanding surplus. Farmland whose value pushed

6. R. Douglas Hurt, *Problems of Plenty: The American Farmer in the Twentieth Century*, 44; Benedict, *Farm Policies*, 115.

7. Saloutos, *American Farmer*, 5–7.

8. Gilbert C. Fite, *George N. Peek and the Fight for Farm Parity*, 9.

9. Saloutos, *American Farmer*, 13.

upward during the war had been purchased at vastly inflated prices by farmers eager to expand their holdings in the face of soaring commodity prices. Rural banks had fueled a 70 percent rise in land values in some areas, as speculation drove farmers to borrow larger amounts. When inflation stopped, the resulting debt burden dragged down the economy of rural America well into the 1930s.[10]

The larger picture reflected agriculture's dependence on the nonfarm economy, the source of much of the 1920s instability. The national business recession that swept the country slowed domestic consumption, and low prices on the international export market hurt staple-crop growers. As historian Gilbert Fite points out, the main problem facing farmers was the disparity between farm prices and the prices of nonfarm goods and services. Between 1909 and 1914, prices between farm and nonfarm commodities were in balance, becoming a benchmark as the "parity" period. The economic disparity following that period was because farmers had no way to control prices that were set by the marketplace and unrelated to the costs of production. The 1920s was a period during which inequalities between farmers intensified. Marginal farmers and sharecroppers—unable to obtain financial capital to improve their commercial situation—remained on poorly producing farms. In contrast, commercial farmers during this period had problems due to bumper crops. Both types of farmers were affected by national recession and unstable international markets, according to historian David Hamilton. The dependence on the nonfarm economy was problematic for an increasingly commercial agricultural industry.[11]

No solution appeared in sight. Although farmers had banded together in cooperative efforts through the Grange and later the Farmers' Alliances, those efforts faded as the wartime nation was urged on to full production for export. The USDA and the federal-state experiment stations concentrated on telling farmers how to increase production, exacerbating the problem of surplus.[12]

10. David E. Hamilton, *From New Day to New Deal: American Farm Policy from Hoover to Roosevelt, 1928–1933,* 10; Van L. Perkins, *Crisis in Agriculture: The Agricultural Adjustment Administration and the New Deal,* 18–19.

11. Fite, *American Farmers: The New Minority,* 31; D. E. Hamilton, *From New Day to New Deal,* 24, 9.

12. Fite, *American Farmers,* 34.

Meanwhile, the world market had changed. U.S. tariffs on imports along with the nation's new status as a world-creditor nation meant other countries had trouble exporting commodities to the United States in order to obtain cash to pay their debts. While the industrial sector recovered, and foreign loans helped alleviate tariffs on manufactured goods, the farm situation remained bleak. The nation tried to have it both ways, to profit from payments by creditor countries while restricting their access to the U.S. domestic market through tariffs. This strategy resulted in high prices for goods that farmers had to buy and low prices for products they sold.[13]

The world wheat situation teetered as weather altered surpluses, providing the only opportunity for profit as more nations entered the world market. In 1925, the Canadian marketing pool successfully dominated the export market as new Canadian wheat crops filled half the global market at good prices. The world crop continued to expand, but prices fell due to a rise in coal rates that slowed trade, creating huge surplus wheat stocks. As the twenties progressed, a series of bumper wheat crops on ever expanding acreage led to swollen stockpiles and falling prices for farmers in Argentina, Canada, and the United States. In 1930, Russia became a major exporter for the first time since the war, flooding the markets with a bumper crop. By 1931, U.S. wheat alone made up 45 percent of the world surplus, as European nations shut out imports to protect domestic growers and growers received subsidy bonuses in Canada and Australia. With a forty million–bushel surplus, even resorting to livestock feeding made no dent in the supply: the world was drowning in wheat.[14]

Responses

The farm problem was central to the national interest, as various groups sought to sustain a landowning class of farmers while adopting the national ideology of efficiency. Solutions were complex and paradoxical, and they often failed. Regional variations in crops, land tenure,

13. V. L. Perkins, *Crisis in Agriculture*, 20; D. E. Hamilton, *From New Day to New Deal*, 5–6.
14. J. S. Davis, Helen M. Gibbs, and Elizabeth Brand Taylor, *Wheat in the World Economy: A Guide to Wheat Studies of the Food Research Institute*, 6, 8.

finance, and marketing created a complex set of problems that varied between neighboring farms as well as between regions. No one solution fit all problems. The twelve-year period following World War I was spent trying to find solutions to the farmers' plight.[15]

Rather than simply extol farmers to raise "two blades of grass in place of one," which farmers had resisted, recognizing it as an effort to move commodity prices further downward, Spillman urged them to develop a strategy for farm survival. He noted the regional differences in farming—soil, weather, and access to markets—suggesting that strategies needed to be regional in nature. No one national model would work for everyone. He advised farmers to follow the practice of the successful farmers in their region, the same philosophy he used to develop his successful extension programs.[16]

The farm public increasingly looked to the government for legislation to increase prices for farm products. Before the war, farmers sought reforms and assistance in the form of expanded credit, cheaper rail prices, and controls on business. During the postwar slump, they continued to seek the same demands, but even credit and transportation reforms did little to alleviate the worsening situation. Farmers prevailed on the government to act in their interests, a shift from the previous view of the government's role as regulator of abuses in the marketplace.[17]

The war had created a federal presence on U.S. farms through the Food Production Act of 1917, which placed agricultural extension agents in every farm county in the country to control and administer farm supplies, credit, and labor, as well as to serve on the local draft board. By 1918, extension agents worked in twenty-four hundred counties across the United States. The Food Control Act (1917) and the U.S. Food Administration under the direction of Herbert Hoover set prices and pushed production. By war's end, farmers were accustomed to federal involvement at the local level and consequently turned to the government for help as they stood on an economic precipice, headed for disaster. After World War I, however, the role farmers, scientists, and the government

15. D. E. Hamilton, *From New Day to New Deal*, 8; Saloutos, *American Farmer*, 14; Christiana McFadyen Campbell, *The Farm Bureau and the New Deal: A Study of the Making of National Farm Policy, 1933–40*; Fite, *George N. Peek*; V. L. Perkins, *Crisis in Agriculture*.

16. Spillman, *Farm Science*, 283.

17. V. L. Perkins, *Crisis in Agriculture*, 20.

would play was not clear. Abundance promised a robust consumer economy, but the paradox of abundance meant farmers experienced both increasing productivity and increasing economic stress. Spillman wrestled with finding a way to balance the situation, initially believing farmers could modify the situation themselves, but eventually advocating governmental intervention.[18]

Two types of agricultural reform emerged, according to historian David E. Hamilton: agrarian interest groups and associational structures. Agrarian interest groups, or "counterorganizations," such as the Farmers' Union and the American Society of Equity evolved to demand higher prices and cheaper credit. They organized to gain power as a group, modeled after professional societies and urban labor unions. Farmers, however, were divided regionally and by commodity groups that ended up competing with each other. A good example is the shifting alliances driven by the margarine industry, as dairy, corn oil, cotton oil, beef, and hog growers found themselves on varying sides of the issue depending upon the market for margarine oils. The alternative to centralized regulation and a strong federal state was the corporatist "associative state."[19]

Many turned to cooperative action as a way to obviate the need for legislative action. Public-private partnerships would facilitate corporatism independent of legislation and regulation. Professional societies, trade associations, managerial elites, technocrats, and scientists sought to create an "associative sector" to maintain social and economic equilibrium to preserve democratic systems without the need of a powerful central state.[20]

Associative reforms were successful in many ways and included the cooperative-extension service (which linked the Department of Agriculture, land-grant colleges, and farm bureaus) and the federal Farm Loan System, created in 1916, which created a base of farmer members-owners. Associational cooperative institutions, such as the American Farm Bureau Federation, provided structure that included many players: farmers, rural business, and government professionals. Herbert Hoover

18. Benedict, *Farm Policies*, 171; Hurt, *Problems of Plenty*, 36.
19. D. E. Hamilton, *From New Day to New Deal*, 4. See also Ruth Dupré, "'If It's Yellow, It Must Be Butter': Margarine Regulation in North America since 1886"; S. P. Hays, *Response to Industrialism;* and Wiebe, *Search for Order.*
20. See Ellis W. Hawley, "Herbert Hoover, the Commerce Secretariat, and the Vision of the 'Associative State.'"

sought to establish national cooperative marketing associations, and the Bureau of Agricultural Economics worked to expand the extension-service model to create local planning arms of a larger national body. The most significant development was the American Farm Bureau Federation, which Spillman had played an early role in helping to organize in Broome County, New York, in 1910. The project began as an effort by the Delaware, Lackawanna, and Western Railroad to revive and maintain a rural customer base. George A. Cullen, a manager at the railroad, contacted Spillman, and with funding from the railroad, the local Chamber of Commerce, the Agriculture Department at Cornell University, and the Office of Farm Management, a local project got under way. A state-college graduate was hired as county agent; he selected a group of community chairmen and worked with the Grange as well.[21]

At first it was merely educational, but as the county extension-agent system expanded, so too did the Farm Bureau. Within a decade, the Farm Bureau had 1.25 million members, mostly in the East and Midwest and a few far western states. Different from the demonstration agent model the GEB had organized across the South, which was a corporatist endeavor between large industry and government, the Farm Bureau involved experts working with an organized group of farmers as well as local business interests.

The Farm Bureaus realized that "no one except a Lenine [*sic*] or a Trotsky would pretend to be able to reform immediately our whole marketing system," but reform must come and must avoid "class-consciousness," according to George Cullen, one of the founders. Cullen pointed to cooperative marketing as the answer: "While fertilizer makes more bushels per acre, co-operative marketing makes *more dollars per man*." Visionary utopian efforts of the past had failed, and it was time to employ specialists to assist with agricultural marketing, Cullen advised. More important, agricultural products had to move beyond local markets, finding the widest distribution possible. That took coordination, expertise, and a coalition among farmers, financiers, and businesses.

21. See D. E. Hamilton, *From New Day to New Deal;* Samuel R. Berger, *Dollar Harvest: The Story of the Farm Bureau;* William Joseph Block, *The Separation of the Farm Bureau and the Extension Service;* Campbell, *Farm Bureau;* Robert P. Howard, *James R. Howard and the Farm Bureau;* and McConnell, *Decline of Agrarian Democracy.*

Bureaus, operating at the local level but tied to a national network, appeared to be the solution.[22]

One reason associationalism did not arise from larger, more established farm organizations, such as the Grange, the Farmers' Union, and the American Society of Equity, was that "all of these are secret organizations, or class organizations, or commercial organizations," explained C. E. Gunnels of the Department of Agriculture in a speech in 1920. "In a considerable degree they are exclusive organizations, and since the work of the Federal and State governments is financed by all the people, these institutions felt the necessity of developing a nonclass, nonsecret, noncommercial" organization.[23]

The Farm Bureaus were not without rural critics, however, as Theodore Saloutos and John D. Hicks have pointed out. Farmers were suspicious, wondering why the railroad and Chamber of Commerce would want to help them and believing that they needed to know not how to produce more but how to get more money for what they already produced. Regional versions of the New York State Farm Bureau phenomenon included the Better Farming Association of North Dakota, an organization made up of business leaders, railroads, and the agricultural college, and the West Central Development Association in Minnesota. Both played significant roles in extending the county-agent system. In the northern plains, the American Society of Equity resisted the associational movements, criticizing the involvement of business interests.[24]

A pamphlet made up of reprints from *American Farming* magazine in 1919 and distributed to Farm Bureau members revealed the social-control aspects of the national organization, which extolled patriotism and nationalism, claiming that farmers, specifically landowning native-born farmers, were organizing as a hedge against fears of bolshevist ideas. Pointing to how labor wielded power, specifically the American Federation of Labor, the pamphlet lauded the Farm Bureau as a way to resist growing urban unrest, attributed to immigrants who "are easily

22. Cullen, "The Cradle of the Farm Bureau Idea: Marketing Possibilities of the Bureau," 2–6, copy in Spillman Papers, MASC.
23. Louis Bernard Schmidt and Earle Dudley Ross, eds., *Readings in the Economic History of American Agriculture*, 557–59.
24. Saloutos and Hicks, *Agricultural Discontent*, 262.

excited to excessive demands and frequently to violence." The pamphlet urged farmers to form a vast organization, to quell discontent if not to allay anxiety. "Just fancy the plight of a Bolshevik or a labor slugger trying to incite riot in a rural district—out in the country where there are no saloons to furnish secretive plotting places, and where the presence of an unemployed stranger who lingers about is regarded with suspicion and distrust!" "Right thinking" in rural America included banding together in an organization that would quell any social unrest emanating from the urban industrial masses. As the pamphlet pointed out, more than 75 percent of the farms in the United States were "cultivated by native-born white Americans. . . . [T]he environment is the very best. There is peace and patriotism in our rural communities, due to regard for properly constituted civil authority. Rioting and violence are practically unknown." The pervasive racist and classist message was directed to more than just farmers, noting that "more than two-thirds of the members of Congress owe their election to rural constituents" and that the growing Farm Bureau movement was "worthy of the serious consideration of all thoughtful people, a matter especially of interest to business men."[25]

The reprinted articles were assembled by *American Farming* under the title "Five Years of the Smith-Lever Extension Act: County Agent Movement," and the cover sported a serious yet dapper image of A. C. True, director of the States Relation Service at the Department of Agriculture. It is hard to tell by looking at the pamphlet whether it was sponsored by the government, provided a promotional device for the magazine, or was of real interest to farmers. It is a clear example of how interrelated the Department of Agriculture, county-level agents, business, and agriculture were as the Farm Bureau movement grew almost exponentially before 1930.

Associations worked to line up legislative support for their constituents in Congress; in 1919, the National Grange hired a full-time lobbyist in Washington, D.C., to represent its five hundred thousand members. County Farm Bureaus organized into the national American Farm Bureau Federation in 1919 and hired a lobbyist in 1920. In 1921, a loosely knit group of nonpartisan congressmen from farm states joined to form the "farm bloc," voting together to protect agricultural inter-

---

25. True, "Five Years of the Smith-Lever Extension Act: County Agent Movement."

ests. Interest-group political pressure grew substantially over the rest of the century but usually focused on short-term benefits—higher commodity prices—rather than long-term planning.[26]

Cooperatives

Marketing cooperatives promised higher prices by avoiding middlemen, something long preached by the Alliances and Granges. Spillman had long been a proponent of farmer cooperatives, an agrarian concept that emerged decades earlier from the National Grange and Farmers' Alliance movement. In contrast to the earlier Alliance movement, postwar farmer cooperatives appeared to be an innocuous vehicle for social control rather than the antithesis to capitalism. In 1918, the Department of Agriculture staunchly supported cooperatives as an antidote to the growing rural dissent. Relying on reports from Russia that the cooperative organizations in Russia had actually been strong supporters of the anticommunist elements there, the USDA saw cooperatives as a useful tool against socialism. The cooperatives in Russia were under central control and had been "persistent opponents of the Bolsheviki and apparently have been largely responsible for many of the civil difficulties encountered by Lenin's government," according to one Department of Agriculture memo. The department believed that cooperatives, under central governmental control, would wreak havoc on any attempts by socialists or communists to overthrow capitalism or the social order. Rural cooperatives, patterned after those in Europe, were accepted by progressives as the solution to both increasing agricultural profits (for landowners) and quelling rural dissent.[27]

Influential Irish reformer Sir Horace Plunkett pushed the idea of farmer cooperatives on both sides of the Atlantic. A friend of Theodore Roosevelt from Wyoming, where Plunkett had speculated in ranching syndicates, he instigated the Country Life Commission and would have been Roosevelt's secretary of agriculture had TR been able to appoint a foreigner. Plunkett worked closely with Theodore Roosevelt

26. Fite, *American Farmers,* 38–39.
27. Unsigned interdepartment memo, September 5, 1918, box 6, file 52, General Correspondence, Records Group 83, Records of the Bureau of Agricultural Economics, National Archives.

and Gifford Pinchot to reorganize U.S. agriculture and rural life along cooperative models while ignoring land reform. Plunkett advocated organizing marketing cooperatives to increase profits for tenant farmers. He saw it as a way to counteract the Irish government's proposed solution to tenant poverty: government programs to subsidize tenants for the purchase of their rented holdings. Higher prices for commodities meant his tenants would continue farming without challenging his position as the largest landowner in Ireland. A fervid anti-Catholic, Plunkett worked closely with Americans interested in stemming Populist land reforms, including Theodore Roosevelt, John D. Rockefeller Jr., Henry A. Wallace, and M. L. Wilson.[28]

Plunkett's attempt to use cooperatives to protect landlords' interests provides an alternative view of the era's widespread support for farmer cooperatives. This contrasts with how historians have viewed cooperatives as self-help vehicles to preserve independence and as grassroots efforts to prevent governmental intervention in the farm economy. Yet both are true. Indeed, rather than becoming an adjunct to bureaucratic control of agriculture, the cooperatives allowed farmers to distance themselves from governmental aid. Farmers were able to challenge "big business" in organized groups that were large scale and efficient. Their group strength created a lobbying interest that also sought to control legislation, rather than becoming dependent upon governmental directives.[29]

Uncontrolled competition was at the heart of the issue, however, as cooperative members saw nonmember farmers profiting from their sacrifice. One aspect that could not be overcome was the necessity for compulsory participation so that nonmembers did not profit from members' sacrifice. Without compulsory pooling of products, a cooperative system could never succeed.[30]

Spillman's Solution

W. J. Spillman had always taken his message to the public. In 1918, when he took the position of associate editor at the *Farm Journal*, he launched a column, "Is Your Farm Sick?" with invitations to "Write

---

28. Daniel T. Rodgers, *Atlantic Crossings: Social Politics in a Progressive Age*, 334–35.
29. D. E. Hamilton, *From New Day to New Deal*, 13; Benedict, *Farm Policies*, 136.
30. John Fahey, *The Inland Empire: Unfolding Years, 1879–1929*, 59.

the *Farm Doctor.*" The magazine introduced Spillman: "For seventeen years he has been connected with the Department of Agriculture at Washington, and for most of that time has been the head of the important Bureau of Farm Management. His work has been to find out why farming did not pay and to show how it could be made to pay. Thousands of sick farms have passed through his hands and, after taking his medicine, have been made well and profitable." This question-and-answer format was very popular in periodicals at the time and allowed Spillman leeway to address virtually any topic he wished. "He knows about marketing, farm bookkeeping, distribution and cooperation—some of the foremost important farm subjects today," the magazine explained. Tempering his qualifications as an expert, the magazine went on to clarify that Spillman was not a "book farmer." "He is a sleeves-rolled-up, simple-hearted, kindly farmer with a twinkle in his eye, and the keen sense of sympathy and humor which goes with it. Everybody likes him and so will you," the magazine claimed.[31]

The *Farm Journal* had adopted a campaign called "A Good Living and 10% for Every Farmer in the Country" as a solution to declining farm prosperity. In the same issue, A. C. Townley's article "Is the National Nonpartisan League the Answer?" explained the newly organizing league and its demand for a "good living and 10%" based on farmers' quest for some way to obtain the value of wages for their labor along with a 10 percent return on their farm investment. Townley, president of the National Nonpartisan League (NPL) and its two hundred thousand members, advocated a political movement to take public control of agricultural marketing. The *Farm Journal* hesitated to give unqualified support to the NPL, but stated that "it is absolutely necessary for farmers to organize, for business purposes, if they are to receive only what they are entitled to, A Good Living and 10%." Spillman's support for the idea was touted to readers: "He believes every farmer is entitled to it, and furthermore, he can help every farmer achieve that result." A sidebar reading "A good living and 10%" appeared on all his columns in the magazine.[32]

Spillman's solution was multifaceted, as were farm problems. He advised planting sweet clover to "cure a sick farm" by improving soil

31. See "The Editor's Viewpoints."
32. Townley, "Is the League the Answer?" 8.

quality. He educated farmers about how to obtain loans from the new Federal Land Banks. Most important, he insisted that the Department of Agriculture provide cost-of-production figures to farmers so they could better understand their economic situation and plan accordingly. Cost-of-production figures were problematic, however, because they could become a foundation for price fixing, with farmers attaching the hoped-for 10 percent profit to any government-generated production figures.[33]

In the 1920s, when questions of national agricultural policy loomed large, Spillman wanted to enlist the public in thinking about how and what science should be applied on the farm. His columns in *Farm Journal* introduced the "new science" of farm management, which "deals with the principles involved in making farming profitable, as well as successful from the standpoint of feeding and clothing the world." He noted that the first lesson to be learned about the new science of farm management was that the "little farm well tilled" was a fallacy. "We have substituted for this doctrine the better one of 'A Good Living and Ten Percent' for every farm family." Spillman's columns were packed with economic facts, practical advice, and a continued emphasis on soil fertility. Touted as "the *man* in Farm *Man*agement," Spillman reached a wide audience, using examples from ancient history, modern scientific applications, and mathematics to make his points.[34]

Spillman was an early advocate for sustainable agriculture as the effects of industrial agriculture began to appear. He wrote that the challenge facing postwar American farmers was how to revive "worn-out farms." He advocated restoring humus by manuring, tilling under straw and crop residue, and planting "green manure" crops such as beans, peas, rye, and clover, which could be tilled under as a soil amendment. But the idea of the "worn-out farm," which had served to gain reader attention for his column in the *Farm Journal,* was already too simplistic. The nation's second-largest crop, hay, and fourth-largest crop, oats, had been sources of energy consumed by draft livestock, which by the 1920s were declining as tractors began to replace them on farms. Spillman noted that by 1918, the market had totally disappeared for six mil-

33. Ibid.

34. Spillman, "Farm Management—a New Science," *Farm Journal,* undated clipping, Spillman Papers, MASC.

lion acres each of corn and oats and eight million acres of hay due to the reduction in numbers of horses. Looking ahead and finding alternatives to the major crops were imperative. Farmers had switched from growing hay and oats to raising wheat and corn, contributing to an oversupply of the latter through the 1920s. Moreover, they were piling on large amounts of commercial fertilizer when they could afford it and adopting efficient techniques wherever they could. Pushing the cost of production down was imperative. Yet as farmers became more efficient, rising production led to oversupply and lowered farm income.[35]

The pursuit of efficiency was becoming a race to the bottom. Just as publications like *Tractor Farming* urged farmers to adopt efficiency as a way to compete in a world market, the efficiency mantra for wartime shifted to become a tool for farmers faced with high-priced land and low commodity prices. An article, "American Farmers! Farm Efficiency Is Essential," told farmers they now faced competition not only from other domestic regions, but from farmers in other countries, too. "It is a generally accepted fact that during the next few years competition between nations will be keener than ever before," and it would be between farmers on an international level.[36]

The Two-Man Farm

Spillman, raised in Missouri, realized that expansive claims for opportunity were not credible under the present conditions. He determined what he believed to be the most efficient farm that also promised an optimum lifestyle: the two-man farm. Influenced by the dire conditions facing cotton farming families across the South, he explained that cotton as a crop limited a family to what the members could pick in one season: no more than six or seven bales. Their income was limited to that amount, which included the labor of women and children. On the other hand, the two-man farm offered a considerably better chance for creating enough income to both support the family and provide some excess capital in order that children could go to school. The

35. Spillman, *Farm Science,* 105, 273; Spillman, *Balancing the Farm Output: A Statement of the Present Deplorable Conditions of Farming, Its Causes and Suggested Remedies* (1927), 44.
36. "American Farmers!" 2.

two-man farm relied on the labor of boys, too, in place of one of the men, but it created more than the income of two one-man operations, and labor could be shared, changing over time as children matured and took their place as "men" in the model. Spillman thought the model gave "satisfactory employment for the working members" of the farm family, but it would be "all the better if the boys can attend school during the winter." Family dynamics for opportunity were part of the family farm, too, as he noted the ideal outcome: "[A] farm of this size [two-man] ought to produce enough income to permit at least one of its young people to extend his or her studies beyond the common schools." Here he drew on his own experience: leaving the crowded Missouri farm to attend college while siblings remained at home and later took over the farm. Viewing higher education as another venue of opportunity, he expounded what Frederick Jackson Turner identified at the time as the next American frontier: the state colleges.[37]

Spillman did not promise Horatio Alger–type success through agriculture; rather, he downplayed the speculative aspect of farming and the quick gain farmers yearned for with a more modest view of success. Being debt free and being a homeowner were clearly Spillman's hallmarks of success, rather than the size of farm or number of men employed. He emphasized that large farms were actually detrimental to the community; scattered population, landless tenants, tenancy—all worked to weaken "schools, churches, and roads," his triumvirate of community structure. His dilemma was to incorporate science into rural life while grappling with ethical issues about the effects of new technology.[38]

Farm Management

Spillman defined "farm management" as being concerned with the business rather than the technical aspects of farming. "It does not deal with how to grow corn," he explained, "but whether to grow this crop, what acreage of it, the cost of growing it, and the returns from it." Farm

37. Spillman, *Farm Science*, 332; Henry C. Taylor and Anne Dewees Taylor, *The Story of Agricultural Economics in the United States, 1840–1932*, 151–52; Turner, "Pioneer Ideals," in *Frontier in American History*, 273.
38. Spillman, *Farm Science*, 335.

management incorporated science, economics, and social science. Proper planning was essential, and the elimination of inefficient, wasteful practices was key.[39]

Farming was evolving from a way of life into a business, a shift most evident in the farm press. The popular magazine the *Country Gentleman* assembled an analysis of content in 1925, aimed at advertisers. It noted that the farmer was

> concerned—and must be more and more as time goes on—with the business side of farming, which means that he must grapple with the same problems that confront every manufacturer. Trade associations, transportation, tariffs, taxes, finance, immigration, foreign markets, politics—these are all matters that have only recently come within his vision, and he needs information, advice, guidance if he is to keep step with other industries in this new era when agriculture is business.

Indeed, farming was becoming a business; that perhaps was the biggest change after the war. As a result, farmers were more interested in "the workings of the economic system, in their relationships to other forms of business and industry, in improved marketing methods, and in a more efficient management of their own enterprises."[40]

Efficiency appeared to be the only hedge or tool on the horizon; it meant reducing labor expenses and speeding up work to cultivate more land. Efficiency equaled mechanization. It was the future. Nevertheless, in the midtwenties, farmers had no cash, little credit left, and few choices to make. Many were enduring drought conditions as well. Spillman shifted his emphasis from soil amendments and selective breeding as ways for farmers to improve their profitability to efficient management.[41]

Facing the new problems of overproduction and a weak agricultural economy in the postwar 1920s, the "worn-out farm" was being replaced by the "worn-out farmer," as people pushed the costs of production down with efficiency. Commercial fertilizers and new strains of seed had moved the issue away from soil fertility as production outpaced

39. Spillman, *Farm Management* (1923), 39.
40. Loring Schuler, ed., *The "Country Gentleman": An Analysis of Content,* 5–6, 8.
41. See Frederick W. Taylor, *The Principles of Scientific Management;* Wiebe, *Search for Order,* 151; and Spillman, "Farm Management—a New Science" (1919).

the market and people worked harder to produce more, to break even as prices fell. Beginning in 1915, Spillman advocated that farmers adopt a system that would spread their labor and energy (draft animals or tractor) use over more months of the year, rather than concentrating on a few months of intensive labor input followed by months of idleness. One way to do that was to develop a diversified farm that included crops and animals that did not all require short, intensive periods of labor at the same time. Field crops, fruit orchards, dairy animals, and forage crops could create a more evenly distributed labor input over a year's time. To survive, the farmer had to adopt the labor system of industry: distributing work over time, with no lax periods.[42]

However, efficiency went only so far; understanding the relationship to the market was vital. Overproduction was key to understanding how the principles of this new science worked, Spillman noted. "Any attempt to stimulate production beyond legitimate demand is ruinous to the farmer. It can benefit the public only temporarily, and that at the expense of the farmer." A concerted national effort at increasing production was irrational, he thought; "the ideal is to increase production as rapidly as is consistent with profit in production." Somehow, production had to be tied to profit. For Spillman, production for its own sake was anathema; profit was essential, and recognizing whether a practice resulted in profit was key. He thought increased production was economically justified in only two cases: when demand increased (due to population growth or better distribution methods) and when the cost of production was reduced. If the cost of production dropped but farmers were still obtaining a margin of profit, they could afford to sell products for low prices, which would stimulate consumption, benefiting both consumer and producer. But if prices *and* profits fell, the demand would increase, spurring additional competition, which would push prices and profits even lower, ultimately destroying producers. "This principle applies everywhere there is free competition and unlimited production," he explained. The rush to the bottom would hurt farmers, the resultant crash would eliminate producers, prices would rise as demand went unfilled, and the boom–and–bust economic cycle would create only negative consequences for all.[43]

42. Spillman, "The Efficiency Movement in Its Relation to Agriculture" (1915), 9.
43. Ibid., 3.

Modern farmers, eager to adopt technology and practices to satisfy an ever changing market, had already abandoned the ideal of the "little farm well tilled." During the antebellum era, Atlantic family farms had been undercut by the expansion into the Mississippi Valley and the adoption of farm machinery. The more efficient western farmers, using new modern methods on larger tracts of land, pushed the Atlantic farms to adopt dairying and intensive farming as swine production and grain farming moved west. The vast West continued to expand opportunity for machinery and larger farms. This "gospel of farm efficiency," based on larger farms and labor-replacing machinery, fostered what M. L. Wilson, a colleague of Spillman at the USDA, called the silent industrial revolution in U.S. agriculture. In its wake, farmers had to either turn to intensive farming for a local market or buy out neighbors in order to expand.[44]

Spillman's idea of farm management was not simply lining up debits and credits and searching for the result. He identified components in farming that could not be commodified. Explaining that the Department of Agriculture's research had found nearly all poultry operations to cost more than the net receipts, he argued that raising poultry nevertheless allowed the farm family to consume eggs and meat of much higher quality than that which was available to the market. The time spent on caring for the chickens should not be assessed as a labor cost because people could do it when they had nothing else to do. Wise use of odds and ends of time could improve one's diet and trim cash expenses for food. He noted that "it is even possible for a farmer to make a fair income when every industry on his farm, if considered separately by the usual methods of bookkeeping, would show a loss." Commodifying a farm family's labor was inherently complicated by the fact they lived and worked on the farm. Keeping books for a small dairy operation nearly always showed a loss, but that was only on paper, according to Spillman. "The fact is that many farmers, if they didn't have these cows, would spend most of the time for six months in the year with their feet against a hot stove." Farm life itself could be an asset or a form of capital, if well managed.[45]

44. E. E. Elliott, "The Farmer's Changing World," 127; William D. Rowley, "M. L. Wilson: Believer in the Domestic Allotment," 277.
45. Spillman, "Efficiency Movement," 11.

Farm Tour

Spillman frequently traveled the nation speaking to professional and farm gatherings, absorbing information about what was happening on the ground. In 1924, he made a nationwide farm tour, visiting Bangor, Maine, and Spokane, Washington, among other stops, to speak before farm and commercial groups. In Maine, he told a Chamber of Commerce group that regional differences mattered—Maine should do what Maine did best: concentrating on feeding cattle, breeding for increased milk production, and cooperative marketing. He spoke to a large gathering of Grange members as well, and the local press covered his talks in depth with daily news stories. Maine farmers were already doing sustainable family-scale agriculture, and Spillman was essentially preaching to the choir during his visit.[46]

During September and October 1924, Spillman made a visit to the inland Pacific Northwest, to the wheat-growing region known as the Palouse, near Spokane, Washington. It was a region heavily dependent on wheat, with climate, market, and transportation challenges to diversifying into other crops. He had left the area more than twenty years earlier, when he resigned as agriculture professor at the Washington State Agricultural College and School of Science in Pullman, seventy-five miles from Spokane. This time he visited thirty-one towns in eastern Washington and northern Idaho on a speaking tour sponsored by the University of Idaho, the Extension Department of Washington State College, the Department of Agriculture, and the Spokane Chamber of Commerce. Spillman's title at the USDA was consulting specialist for the Bureau of Agricultural Economics. "Dr. W. J. Spillman . . . a world authority on agricultural economics, was the leading feature of the tour," the Chamber of Commerce noted. A large automobile entourage formed, including Spokane bankers, representatives from the Northern Pacific Railroad, and representatives from the agricultural newspapers. The basic theme was diversified agriculture, which Spillman touted as the route for successful farm life in spite of the meager financial returns.

46. "Local Farm Tour Features Expert," *Bangor (ME) Daily Commercial,* March 7, 1924; "Dr. Spillman Discusses Farming with Grangers," *Bangor Daily Commercial,* March 8, 1924.

Listeners had reason to turn out; cash returns from wheat, the region's mainstay, were low.[47]

Many aspects of farm life in the region as well as the nation had changed in the thirty years since Spillman's wheat genetic experiments in eastern Washington. In 1894, however, farmers were already planting large acreages; farms between 1,800 and 2,500 acres were not uncommon. Wheat crops had been big, and the risks had been relatively small. Although the crash in 1893 had lowered wheat to $0.25 a bushel, by 1902 one farmer reportedly sold his crop for $54,000. Labor was inexpensive, about a dollar a day for farmhands and $1.50 to $3 per day during harvest and hay season. Raw land was $5 to $7.50 per acre, and expansion was easy. Joe Smith, a local resident, remembered his father's farm during that period. His father figured one crop of wheat could almost pay for a 160-acre farm, and the second year's crop was free—growers let the second year's crop come up voluntarily after grazing cattle on it over the winter. One year the volunteer crop had better yields than the original crop planted the preceding season. Letting the previous year's wheat crop reseed itself while tilling up a fresh section of land for planting was economical, as long as an ample supply of land remained available. The supply of new land quickly disappeared, so farmers increased investment in mechanization when possible. Nevertheless, the region had few local markets, and a network of rail lines reinforced wheat exports as the major crop. In the heyday of wheat growing before and during the war, the region's farmers had prospered. Now with the global market glutted and prices below cost of production, growing wheat no longer guaranteed prosperity.[48]

Spillman's gift for public speaking was evident, as he told anecdotes about growing up on a Missouri farm without cash. "We did without most of the things" that required money, he admitted. He preached only one message to the economically depressed farmers: balanced farming. He admitted the local cash crop—wheat—was central to the region, much like the Midwest's reliance on corn and the South's dependence on cotton. He advised families to move away from "one-crop farming"

47. Spillman, *Balanced Farming for the Inland Empire: The Story of Dr. Spillman's Lectures during the Agricultural Tour of 1924* (1924); Fahey, *Inland Empire*, 85.
48. Smith, *Bunch Grass Pioneer*, 15–18.

because monoculture was detrimental to the soil; seasonal idle periods encouraged farmers to take jobs away from home, weakening the family; and relying on a single commodity was unpredictable. He said one-crop farming was "hazardous in any section and in most sections pre-doomed to failure." Instead, he encouraged farmers to improve their soil through a mix of livestock, forage crops, home orchards, and gardens. Balanced farming meant incorporating many (but not all) of the following into a farm that enriched the soil, ensured a healthy lifestyle, and brought in some annual cash income: cattle, pigs, sheep, chickens, horses (as draft animals), alfalfa, clover, wheat, barley, oats, corn, fruit trees, bees, or other appropriate crops.[49]

As he proselytized for balanced farming, he used one of his typical folksy proverbs to encourage farmers to accept the paradigm he advocated. He used a rabbit story to encourage farmers to shift from monoculture: "Bunny was hard pressed. Another weary leap or two and he would be overtaken by a pair of gaping, ravenous jaws. To escape them, he turned aside and climbed a tree. 'But a rabbit can't climb a tree,' interposes the listener. 'Yes, I know a rabbit can't climb a tree,' says the story teller, 'but this rabbit had to.'" Pressing the wheat growers, he told them that "if rabbits can climb trees when they have to, wheat farmers can change their system of farming, when they have to. And they have to."[50]

He gave them two main reasons to opt for diversified farming: first, continuous wheat cropping had depleted the soil, lowering yields, which made the cost of production higher; and second, the market for wheat at "living prices" was limited, leading to frequent years of diminished net incomes. He reminded the audience that when he taught at the state college in Pullman twenty-five years earlier, Palouse farms produced as much as sixty bushels per acre, with averages of forty to fifty bushels per acre. He contrasted those figures to the current yields that were running about twenty-five bushels per acre on average, with forty-five bushels under only perfect weather conditions. Summer fallow, the practice of letting land lay idle every other season, followed by wheat, year after year, was diminishing the soil's fertility, a process Spillman likened to "drawing on the principal in the bank account." Using the

---

49. Spillman, *Balanced Farming*, 6.
50. Ibid., 7.

metaphor of capital to describe the mineral and organic nutrients in the soil, he explained that present farming practices were drawing on the soil without replenishing nutrients. Eventually, the farmer's asset— soil—would be wiped out. He compared the diminishing yields to a decreasing bank account—the end result would be "insufficient funds."[51]

Whereas diversification meant stability, the farmers' immediate concern and the one they were most interested in was the market for wheat. Could they expect those boom years during the war ever to return? Spillman noted that the world population was increasing, but so was international wheat production, so demand was stagnant. Overproduction was here to stay, and prices would continue to fall as more land worldwide went into wheat, he advised. The postwar world was experiencing much the same thing as the United States: draft animals were being replaced by the internal combustion engine, diminishing the need for forage crops that had been marketed locally. As forage acreage was replanted in other crops, farmers were forced to plant commodities that could be marketed at a distance: corn, wheat, and cotton. As petroleum replaced oil seeds for fuel and illuminants in Europe and Asia, acreage once planted to oil seed crops was being planted to marketable commodities, too.[52]

On a global level, supply and demand would never let prices rise much above the average cost of production. "In the case of a world commodity like wheat, the price tends to fluctuate about the average world cost of production," Spillman explained at every stop. "If the price goes much below that level it will reduce production in those regions in which the cost of production is highest and this will tend to bring the price back to normal." Whenever prices rose above the average, farmers would quickly increase production and prices would drop. This cycle would be disturbed only by shortages due to insects or weather, when normal supply would fall off and therefore prices increased for a short term. In 1924, Europe's shortage had led to a price surge for U.S. growers, but it was short-lived and unlikely to reoccur because Argentina and the United States had increased production the following year. Farmers were feeling that 1925 glut as he spoke.[53]

51. Ibid., 8.
52. F. H. King, *Farmers of Forty Centuries; or, Permanent Agriculture in China, Korea, and Japan*, 190–92.
53. Spillman, *Balanced Farming*, 9.

Spillman detailed the effects of production in the four wheat-exporting nations: the United States, Argentina, Australia, and Canada. In the past thirty years, the United States had increased wheat production from 643 million bushels to 838 million bushels. Argentina had more than doubled its production, Australia had trebled its production, and Canada had nearly doubled wheat production in the past two decades alone. After describing how the four countries had increased production, he noted that the tendency would only continue. He told wheat growers their only hope to profit was to become more efficient at wheat culture. Lowering the cost of production was the only way they could hope to stanch failure in light of diminishing prices. To lower costs, they had to increase production without increasing their acreage. Spillman advocated soil fertility as the way to achieve more from the same inputs.

Soil fertility symbolized a rich and healthy farm life, something he thought could be accomplished by balanced farming. Livestock on the farm were key to both improving the soil and providing additional food and revenue sources for the family. He thought farm families should look for a variety of cash income sources, telling about one woman's claim that her family had bought all groceries and clothing with income from chickens, butter, hogs, and garden sales, reinvesting their wheat check into growing wheat. Livestock remained essential; they were what Spillman called "the keystone in the arch of balanced farming." Livestock as an adjunct to wheat growing would earn farm families a better life as well as additional income, while continuing to allow them to grow wheat, which the region's farmers and bankers were not about to give up. Raising livestock would lessen soil erosion, too, he noted. The region's soils had been washing away, visible in the gullied hillsides all around them, which grazing might have prevented. Leaving steep and vulnerable areas in grass for grazing would return humus to the soil, bind the soil particles together, and increase moisture absorption, he claimed. It would help prevent the loose topsoil on the Palouse hills from washing away.

Livestock manure was not the mainstay of Spillman's philosophy, even though it did replenish the soil. Rather, he saw that maintaining livestock could reshape the entire cropping system. Getting farmers to raise more livestock on the farm meant a need for more forage, something that would silently diminish the supply of unsold wheat. He advocated planting clover, alfalfa, peas, and beans, which, in rotation with wheat, would both provide forage for animals and enrich the soil with the nitrogen-fixing action of the roots and when plowed under also provided soil amendments. He did not stress that incorporating these other crops and livestock into a regional cropping pattern would mean fewer acres planted to wheat, which would reduce the overall supply.

Spillman's scientific advice was skillfully tendered with selfdeprecating common sense, as when he described difficulties he had faced getting cattle to eat the clover he had grown. Woven into his tales of recalcitrant cattle, he launched into explanations of legumes, nitrogen-fixing bacteria, and soil "inoculation." Inoculation would supply needed bacteria in the soil that transmitted nitrogen from the air to legume roots. Seeds could be inoculated by treating them with laboratory bacteria cultures or by sprinkling soil that contained active cultures onto the field when planting. Legumes' extensive root systems also aerated the soil and decomposed as humus.

Livestock encompassed more than beef cattle, however, which were in a postwar price decline. Spillman urged sheep growing as one alternative because wool markets promised to expand, whereas supply was limited. U.S. textile factories were importing two-thirds of the wool they used—a market domestic growers should capture. The only risks would be if the wool import tariff declined or a wool substitute developed—both he deemed unlikely. Dairy cattle were another possibility, but he realized wheat growers disdained the level of labor a dairy herd required. Nevertheless, it was an option that provided profits when wheat prices were low. Aspects of dairying went beyond profits, because milking cows provided a consistent year-round return on labor. The farm family would be tied down to caring for the animals, but family members would not have to relocate to the mines or logging camps over the winter, disturbing home and community life. Dairy cows kept the family at home, centered on "a permanent type of farming, a more studious and thoughtful type, a more self-sufficient home type." Horses, a longtime staple animal in the region, were no longer a viable source of income, he advised. Demand was falling and would continue to do so, as the wartime demand for military horses disappeared. The need for horses for pulling farm implements was fading, and the urban market—horses for pulling all sorts of equipage—was being replaced by electric streetcars, automobiles, and trucks.[54]

As farmers were abandoning livestock for mechanically harvested crops that could be grown without much manpower, progressive-minded farmers might have viewed Spillman's advice as backward. Livestock, in many farmers' minds, were not the future—they were the past. Machinery, tractors, the gasoline engine—the future was in mechanization. To suggest that growth, expansion, and prosperity might not continue naturally forever "meant in most communities to be branded as a dangerous eccentric." Worse, it clashed with the "speculative and commercial tendencies that were growing up in modern American agriculture," according to E. E. Elliott, an agricultural economist with the Bureau of Agricultural Economics in 1940.[55]

It was hard for Palouse wheat growers, and indeed the rest of the nation's farmers, to accept what Spillman was saying. They had just experienced a tremendously profitable period, and to argue it would never return contradicted the national agricultural myth. "Boomer psychology" characterized how Americans experienced the idea of progress. The ideology was "one of the most important of all the influences that have shaped the course of American agricultural development," according to Elliott. Rising land values and increased settlement had created a boom-bust cycle since the eighteenth century. American farmers expected hard times to be followed by cycles of opportunity.[56]

To area residents, particularly the younger generation that was taking over their parents' land, farming was a capital-intensive endeavor, not a make-work life of taking care of animals, gardens, and milk pails. As the *Grange News* observed, "The average wheat farmer will not milk cows." Wheat farmers were heavily in debt for expensive equipment and land and unable to risk changing crops when no clear markets existed. John Fahey, a Pacific Northwest historian, points out that although the farm audiences respected Spillman, they might have considered his advice old-fashioned. Nevertheless, without a market, continuing to grow wheat was clearly the wrong path to take. Spillman's

---

54. Ibid., 18.
55. Elliott, "The Farmer's Changing World," 129–32.
56. Ibid., 132.

balanced farming required more than a balance on the farm; it also encompassed a balance in the market. Efficiency alone could not adjust production to market demand.[57]

Balancing the Farm Output

The Palouse farm tour changed Spillman's thinking; he returned to Washington, D.C., determined to find a solution to the farm crisis. Improved breeds of plants and animals, soil amendments, farm credit, efficiency, and business practices had not helped farmers maintain a secure position. Cooperative marketing, though helping ameliorate the situation for some, had not lifted all boats. Science, efficiency, and cooperation had failed to resolve the growing dilemma of too much production for market demand. Spillman believed that there had to be a solution, a way to "adjust the production of farm products to market demand in such manner as to enable the farmer to get a fair price for his products." He used the term *balance* to suggest what needed to occur to adjust production to market demand. In a new book, *Balancing the Farm Output*, he laid out his new ideas for adjusting farm production to market demand while maintaining a fair price for farmers. In the preface, he thanked Secretary of Agriculture William Jardine for "permitting the publication of the volume with no limitations whatever on its contents." His earlier experiences with departmental censorship during the years under Secretary Houston were behind him.[58]

Spillman noted that, in 1927, surplus planting of wheat beyond market needs was nine million acres and cotton was ten million acres. Mechanical power was part of the problem, causing a 50 percent decline in the number of horses on farms. Tractor use had expanded from simply plowing to cultivating and working in row crops like corn and cotton. Previously, corn growers had limited production to eighty acres—all that two men could work without a tractor. But tractors removed the greatest limitation to farm expansion: lack of labor. Spillman saw that

57. *Grange News*, July 5, 1929, cited in Fahey, *Inland Empire*, 85; Spillman, *Balancing the Farm Output*, 12.
58. Spillman, *Balancing the Farm Output*, 12.

while the tractor was enlarging farm acreage, it was reducing the number of laborers, reducing rural populations. By 1927, the situation created a surplus of thirty-nine million acres planted to commodities that had no market. The amount of land planted in the nation's "big five" (corn, wheat, cotton, oats, and hay) was so immense that a comprehensive effort was imperative. He noted that the "twenty-two million acre reduction in the area of the wheat crop since 1919 was brought about largely by the abandonment of farms." The growing reserve of dormant cropland loomed over any attempt to adjust production because any rise in price would put the land back into wheat, dropping the price back below the cost of production. To prevent that, Spillman advocated a multipronged effort to take land out of cultivation, through planting forests, creating permanent pasture (which could be planted back to crops if the market increased), and increasing livestock (sheep, dairy cattle, beef cattle, and hogs), particularly in the Corn Belt, where animals would consume excess corn.[59]

Spillman figured statistical information might be another solution. The Department of Agriculture issued an *Outlook Report* in February of each year and predicted what increase or decrease in planting might work to improve prices based on that commodity's normal yield. The *Intentions Report* followed in mid-March and told what farmers planned to do. The statistical information had been useful to bankers and commodity dealers, but growers had little opportunity to access it. Spillman wanted to expand dissemination of the statistical information about crop projections so farmers could apply that information during spring decision making. Knowing what everyone else had decided to plant (or not plant) allowed farmers a final opportunity to modify harvest outcomes at planting time. Information in the form of statistical information about production and marketing might be the farmer's salvo. He called it "The Plan" and thought it was bold and simple enough to succeed at putting statistical information in the hands of people just as they made crucial decisions about planting. It would allow farmers to stabilize production of minor crops where seasonal fluctuations in production directly affected profitability. He mapped out a plan for farmers to gather at local schoolhouses to elect officers, who would work

59. Ibid., 45.

with the county agent to examine each winter's *Outlook Report* issued by the Department of Agriculture. Six weeks later, the farmers would receive the department's *Intentions Report*, and they would meet again to discuss it. As a local group, they would consider whatever regional crops or livestock were most important to them. It would be democratic, local, and cost free.

But he also realized that it was impossible for cooperative efforts by farmers to limit their acreage, no matter how much statistical information was available to them. They could not create a farm trust or a livestock trust; those efforts had failed. Growers could not pattern their industry after the business model based on creating trusts to control production and marketing; that model had worked only for marketing regional products, such as the Sapiro cooperatives in the California fruit industry. Using information from the *Outlook Report* and forecasts to shape production would likely fail because psychology and competition drove farmers' decisions about planting. Farmers still gambled on what and how much to plant.

Spillman argued that a laissez-faire agricultural policy would never put producers on an economic basis similar to that in other industries because the amount of available land that could be put into production when prices improved would always keep prices low. He argued that "it is generally recognized that the present economic situation in the United States works to the disadvantage of farmers and of the salaried classes." He looked to what he called "government interference" as the solution. How government would intervene was debatable.[60]

Much of what Spillman addressed was opposed to Herbert Hoover's agricultural vision. Hoover believed that a cadre of professionals, managing from the top down, was necessary to control farmers because they were a threat to self-government. He abhorred waste, exalted efficiency, and strongly supported tariffs as a "lever" for production adjustments. Hoover's vision was one of cooperation between managers and middlemen, ignoring the farmer's role. Hoover epitomized agricultural management by a handful of professionals, in a top-down manner, whereas Spillman advocated disseminating information to farmers (such as in the *Outlook Reports*) and fostering action in independent local groups.

60. Ibid.

Hoover's election in 1928 secured passage of his Agricultural Marketing Act the next year, but farmers and legislators alike only weakly accepted it. Although it was a landmark of sorts, the first step toward federal involvement in bringing agriculture and industry toward economic equality, it was largely ineffective.[61]

Proponents of stiffer tariffs on surplus products for export believed farm products could be protected similarly to protections in place for the manufacturing industry. Spillman explained that protective tariffs, imposed on agricultural products such as wheat (forty-two cents per bushel tariff on imported wheat), had little effect because imported wheat was not holding down domestic prices: little wheat was imported at all. He pointed out that agricultural products could not be treated like manufactured goods, which could be sold at higher prices to a domestic market while exported at lower prices outside the country because they had no foreign competition. "The exportable surplus of such products does not control the price of that portion of it consumed in this country," he explained.[62]

Spillman pointed out that even though many farmers insisted on a plan to facilitate selling agricultural products at one price on the domestic market and at another price on the export market, it was impossible due to foreign competition. Again, manufacturers could operate that way, if they had no foreign competitors, but farmers could not.

Wages were under attack, too, as foes of labor union organization claimed high wages for labor were responsible for the farmers' woes. Spillman thought the high wages were a "fortunate thing," and although they drove up labor costs and the price of goods farmers had to buy, he believed the farm situation should not be used to justify forcing wages down "to equalize the economic status of agriculture and industry." He noted that labor union leaders wanted farmers to attain a larger share of the nation's income, too. Spillman stressed that the increased buying power of wage earners had boosted industry and that a rise in farmers' income would also "be a very strong stimulus to industry in general." He worried that legislation to limit wages might be "forced on the country," which would devastate the economy. He advocated solv-

61. D. E. Hamilton, *From New Day to New Deal,* 34–35, 39.
62. Spillman, *Balancing the Farm Output,* 62.

ing the farmers' income inequity, calling it "a patriotic duty" to resolve the situation, while preserving wage gains that bolstered the consumer market.[63]

Spillman considered the wide range of remedies proposed to resolve the farm dilemma, including: production bounties, export bounties, price fixing, contracting products, and valorization. All were flawed. Production bounties would only increase the surplus in the subsidized crop. Export bounties would bring better prices but would be quickly met with increased production and resulting flooding of the market. Price fixing would require decades, until the market grew to meet production capability, and with the mass of untilled land that could be put into production, that would be far in the future. Government valorization (establishing and maintaining a set price that provided a profit) would only spur production in other countries, and the nation would lose an entire industry. Any effort to create higher world prices would only stimulate competition from other countries that could grow the same commodity. Competing countries had wiped out the coffee industry in Brazil, the same with Britain's rubber plantations and the sugar industry in Cuba. The futility of any type of effort to decrease production would always be met with the huge amount of land that could be quickly put into production, particularly with mechanization.

Spillman argued against simply subsidizing exports, noting that export subsidies paid by the government would not solve the problem because the United States already filled one-third of the world wheat market of six hundred million bushels and could easily expand production by four hundred million bushels. Increasing the amount of domestic wheat on the world market would only lower the world wheat price with an oversupply. The nation's problem was not an oversupply of farmers, but rather an oversupply of *land*. Spillman pointed out that *"we have so much idle land waiting to come back into cultivation"* that any government program had to be certain *not* to provide any incentive for bringing more land into cultivation. "There could be no compromise whatever on this point if the plan is to be successful," he advised.[64]

63. Ibid., 64–67.
64. Ibid., 74, 97. Italics in original.

Spillman's Plan

Spillman presented a plan that he believed was "economically sound, that would increase the price to the producer by any reasonable amount desired, would not stimulate increase in acreage, and would not in any way affect the world price of the commodity." He called it the "Limited Debenture Plan" and patterned it after the procedure milk dealers in Baltimore and Philadelphia had adopted to stabilize milk production. Each milk producer held an allotment at an agreed-upon price; any milk supplied over that amount was purchased at the market price, which was usually lower. The "Limited Debenture Plan" Spillman envisioned was an allotment plan that would be based on a farmer's average production and would guarantee a profitable return on the allotted acreage, but would also allow the farmer to sell additional crops at the market price. The farmer would receive a support payment equal to the domestic market share of his average production, set when the program began. A target price would be set for the crop, paying the support subsidy only when the market price fell below the target price. Managing the land, not the farmer, was the solution. Spillman thought the allotment should reside with the land, not the farmer, if it changed hands. He realized that it was the *land* that could be controlled and managed—not the farmer. Applying government programs to acres did not threaten independence and democracy. People were not bound to the soil; it was the soil that could be bound to the state. In one swoop, the nation's cropland would be nationalized, yet farmers would continue to live as yeomen.[65]

But the Limited Debenture Plan would not resolve the sociological problems farmers faced, particularly the diminishing opportunity for those without capital or on poor land. As mechanization and capitalization became more significant on the farm, economies of scale still promised a better return than the "small farm well tilled." Spillman admitted that his plan would not "retard the industrialization of farming...unfortunately." He pointed to the paradox that "anything that tends toward making farming profitable will tend to hasten the industrialization of it." He described industrial farming as concentrating in

65. Ibid., 82.

large units "for economy of production, but at the same time making the vast majority of farmers dependent laborers instead of independent operators."[66]

In 1933, Franklin Roosevelt was inaugurated as president and appointed the young Henry A. Wallace as secretary of agriculture. Wallace, with the help of Milburn Lincoln Wilson, a professor of agricultural economics at Montana State Agricultural College, brought Spillman's Limited Debenture Plan, known as an allotment plan, to Roosevelt's attention. Wilson had been a believer in efficiency and mechanization on larger farm units until the failed economy of the 1920s pushed him to adopt a different ideology. After working with Spillman at the Office of Farm Management from 1924 to 1926, he absorbed Spillman's economic philosophy and eventually became a major proponent of his ideas. Wilson admired Spillman but was at first unswayed by his balanced-farming ideology. In Montana, Wilson administered the Fairway Farms Project, a Rockefeller Foundation program operated through the state college to buy up wheat farms, converting them to large-scale mechanized farming operated by tenants, while studying the most profitable means of efficient production. He continued to believe in the doctrine of increased efficiency and production until he realized that most of the investment in mechanical farming was coming from manufacturers, not farmers. He was so entrenched in large-scale mechanized farming that the Soviet government hired him as a consultant for large-scale tractor farming there. Even though he realized that Russian wheat would soon swamp world markets, he continued to believe efficiency would save Montana growers, until he could no longer obtain financing for large-scale projects from the Rockefeller Foundation in 1929. Shortly after, Wilson became an advocate of the limited production plan, largely based on Spillman's work.[67]

Ironically, in light of Spillman's longtime conflict with Rockefeller Foundation funding in the USDA, it was the Laura Spelman Foundation, a branch of the Rockefeller organization named after John D.

---

66. Ibid., 110.
67. Ibid., 277, 281; Rowley, *M. L. Wilson and the Campaign for the Domestic Allotment,* 48; Rowley, "Wilson: Domestic Allotment," 277, 279.

Rockefeller's wife, that helped spread Spillman's allotment theories. In 1929, John D. Black, a Harvard economist, with funding and direction from Beardsley Ruml, a psychologist who headed the Spelman Foundation, wrote a study, *Agricultural Reform in the United States,* based on Spillman's ideas. The book was targeted at the incoming Hoover administration and sought to gain support for a domestic allotment plan. It is not clear why the Rockefeller Foundation was eager to support the idea, but Ruml was specific about not being identified with Black's book and that it not be pushed as a "plan." With advice from Chester Davis, a lobbyist, and Henry Taylor, former Bureau of Agricultural Economics head, Black worked with Ruml to get the book into the hands of congressmen, bypassing the farm associations. M. L. Wilson read Black's book, translated the ideas into what he termed the "Domestic Allotment Plan," and eventually joined Henry A. Wallace in introducing the idea to Governor Franklin D. Roosevelt, as Roosevelt's presidential campaign took off. As Wilson promoted the domestic allotment proposal, he avoided connecting it to Black and Ruml. Instead, only months after Spillman died, he began touting it as Spillman's plan to farm groups who were suspicious of Wilson and his supporters. Wilson promoted the plan to the Grange, Farm Bureau, and Farmers' Union by identifying it with Spillman and "portrayed himself as the intellectual heir of the plan after Spillman's death," according to historian William Rowley.[68]

The plan brought farm groups such as the Grange and Farm Bureau together with Roosevelt and the Democrats, perhaps because Wilson linked it to Spillman, whom the groups respected. Finally, all parties seemed agreeable on the direction to proceed. Historian Christiana McFadyen Campbell notes that it was perhaps the last opportunity. She quotes George Peek, an earlier proponent of governmental intervention to control the surplus, as saying, "It looks to me as though in the campaign for Roosevelt . . . we are in the last trenches and if he is not elected that agriculture is doomed to peasantry."[69]

Ramsay Spillman noted the irony of the Rockefeller involvement in gaining acceptance for what is considered his father's most important

68. Saloutos, *American Farmer,* 35–36; Dean Albertson, *Roosevelt's Farmer: Claude R. Wickard in the New Deal,* 50; Rowley, *Wilson and the Campaign,* 105.
69. Campbell, *Farm Bureau,* 51.

contribution to agricultural economics, by sponsoring publication of Black's book. "It would have appealed to my father's incorrigible sense of humor to witness the Rockefellers financing the publication of a book of which an important portion was the modification of an idea that was evolved in the brain of the man who wrote the legislation which removed the Rockefeller funds from the Department of Agriculture," Ramsay wrote.[70]

In 1931, Spillman died from an infection following surgery for chronic gallbladder problems and was unable to play a part in crafting the agricultural New Deal, a subject he had spent years formulating. Writing only months before he died in 1931, he realized the extent of the crisis and the futility of letting the agricultural situation resolve itself naturally. Finally, however, he had identified the cause of consistent overproduction. He pointed to technology as the root of farmers' woes since the close of the Civil War. He explained that the Greenbackers and Populists had been only partly right; the nation needed additional infusions of capital in order to finance the increased agricultural production, but "the big financial interests of the country" they had blamed were not as significant as technology. He chronicled the adoption of five innovations: the steel moldboard plow, mower, reaper, threshing machine, and two-horse cultivator. They had greatly increased the amount of land a family could work. That technology, along with the opening of government land in the West, led to overproduction and low prices by the 1890s. Recovery and wartime had intervened, but by 1920, overproduction again reached crisis proportions, as land shifted into wheat and cotton, glutting a declining market. Technology had altered the situation again, as the tractor, truck, automobile, and combine increased the size of family farms. Planting technology and genetics fostered wheat and corn farming in previously arid areas. However, the same had occurred in Canada, Argentina, and Australia, and Russia's large grain production was rebounding.[71]

Balanced farming—diversification based on soil fertility and livestock—had not lived up to the promise Spillman had once envisioned.

70. R. Spillman, "Biography of Spillman," MASC, 392.

71. "Dr. W. J. Spillman, Scientist, Dead," *New York Times*, July 12, 1931; R. Spillman, "Biography of Spillman," MASC, 346; Spillman, "Agricultural Crisis" (1931), Spillman Papers, MASC.

The turning point for Spillman was evidenced in his 1927 book, *Balancing the Farm Output*, where he turned from independent solutions, such as diversification, to the state. "Without governmental interference of any kind, it would appear that the crisis must continue and possibly grow more severe," he predicted. Agriculture was indeed in a crisis, and the only alternative was federal intervention.[72]

The Limited Debenture, or allotment, Plan, appealed because it limited production but was enforced at the local level by an elected county committee. It was a politically feasible way to control production. Resistance came from milling companies and grain dealers, as well as the Chamber of Commerce, which saw no need to help agriculture, and Herbert Hoover, who resisted government farm programs, relying instead on managed laissez-faire.

In 1932, Franklin Roosevelt's election signaled a new era, and within six months of the enactment of the Agricultural Adjustment Administration (AAA) in 1933, millions of dollars had been funneled to growers, half the nation's farmers had signed on to federal programs, and long-term reduction of production in major commodities was in place. Four thousand local committees administered federal programs, "making it the most decentralized and democratically participatory federal program in the nation's history," according to Henry Wallace biographers John Culver and John Hyde. Roosevelt admitted, "I tell you frankly it is a new and untrod path, but I can tell you with equal frankness that an unprecedented condition calls for the trial of new means."[73]

The AAA was the first New Deal recovery program, and as other programs were put in place, including a controversial initial period of plowing under growing crops, dumping milk, and slaughtering baby pigs to restore supply to market demand, Roosevelt's plan eventually succeeded. They "were not acts of idealism by any sane society," Henry A. Wallace admitted, but were "emergency acts made necessary by the almost insane lack of world statesmanship during the period from 1920 to 1932."[74]

In 1935, Secretary of Agriculture Henry A. Wallace spoke at the Iowa State College, explaining that he had met Spillman twenty-five

72. Spillman, "Agricultural Crisis," MASC, 4; Rowley, *Wilson and the Campaign*, 37.
73. Culver and Hyde, *American Dreamer: The Life and Times of Henry A. Wallace*, 125; Rowley, *Wilson and the Campaign*, 177.
74. Culver and Hyde, *American Dreamer*, 125; Wallace, *New Frontiers*, 200.

years earlier, when Spillman's "Messianic complex about the farm-management approach" had reshaped Wallace's thinking. "Prof. Spillman" had urged him to "go out on the farm and get first hand knowledge," which hardly seems remarkable now, but as Wallace explained, it was contrary to the "Aristotelian" training agriculturalists and economists were getting at the time. Wallace, secretary of agriculture from 1933 until 1940, noted that Spillman was "the philosophic father" of the Agricultural Adjustment Act. Indeed, Spillman's "plan" to limit production through agricultural allotments based on average production, and tied to the land and not the farmer, molded and shaped national agricultural policy for the rest of the century.[75]

75. Wallace, *Democracy Reborn*, 96.

6

Conclusion

Farm Evangelism and Dollar Efficiency

In 1931, William Spillman suffered from what his son, Ramsay, described as a chronic infection of the gallbladder that impaired the pancreas, causing a diabetic cataract in one eye. By early summer, his vision was gone in that eye. Opting for gall bladder surgery, he urged Ramsay to be sure to take care of a manuscript on calculus he had prepared but that had not been published. Later, Ramsay learned his father had dedicated it "to my son, Dr. Ramsay Spillman, whose difficulties with calculus during his academic days led me to undertake the preparation of a simplified treatise on the subject." He called it "A Simplified Calculus" and wrote it as a university textbook.[1]

Ramsay, a practicing medical doctor, believed his father died from peritonitis as a result of the surgery. A postoperative infection could not be contained in the days before antibiotics. Realizing his condition, William Spillman spent two days in the hospital before he died on July 11, 1931.[2]

There were several memorial services, one initially for family and friends in Washington, D.C.; another that autumn in Pullman, Washington; and a dinner that December at the meeting of the American Farm Economic Association in Washington, D.C., at which 120 friends

1. R. Spillman, "Biography of Spillman," MASC, 293. Two unpublished calculus manuscripts are in the William J. Spillman Papers in the Western Historical Manuscript Collection, University of Missouri–Columbia Archives.
2. Ibid., 376.

and associates remembered his professional achievements, including his role as the founding president of that organization.[3]

The memorial service on the Washington State College campus, where Spillman began his scientific career, was well attended. Representatives of the governor of Idaho, University of Missouri, USDA, Washington State Grange, and Washington State Farm Bureau; editors of farm magazines; and more than a dozen regional Chamber of Commerce delegates attended. It was an era of memorials for men who had contributed to scientific education, notably David Starr Jordan, whose passing had been marked at Stanford University the month before, and the recent death of Thomas Edison, which had also been marked with tributes. In Pullman, college officials planned a suitable service for Spillman, who many claimed had touched their lives and distinguished both himself and the agricultural science community. As Enoch Bryan, the main speaker and retired Washington State College president, noted: "It was here in this state and in this Pacific Northwest that he first endeared himself to the farmer in such a way that throughout the United States he was often looked upon as a Moses who might lead them out of the wilderness."[4]

Bryan identified the crucial difference between Spillman and other scientists or farmers: he had a solid grounding in science, hands-on experience in agriculture, and the "inventive turn of mind" needed to develop courses of study that both satisfied the needs of practical agriculture while challenging students with sophisticated subject matter and pedagogy. He also had a gift for speaking to popular audiences and a keen sense of humor. After Bryan offered him the position as agriculturist at the new college, Spillman threw himself into the career with gusto. "He knew at once what his life work was to be," Bryan recalled, "and he never doubted nor faltered until his untimely end." An omnivorous reader and autodidact, Spillman was personable and sociable, able to pass on the knowledge he gathered. That had been crucial to developing a link between scientific research based at land-grant colleges and the public. As Bryan noted, Spillman realized that the success of the

3. See T. N. Carver, H. C. Taylor, and G. F. Warren, "William Jasper Spillman, 1863–1931: First President of the American Farm Management Association." The organization changed its original name to the American Farm Economic Association.

4. Bryan, memorial address, October 22, 1931, Spillman Papers, MASC, 13.

land-grant college movement depended upon public contact. "He must become a missionary of the gospel of a new agriculture among the very people for whom it was chiefly intended," Bryan explained. Spillman, indeed, had molded himself into what Bryan described as a "farm evangelist," promoting the positives about farm living, honoring tradition, and always seeking ways to make farming a satisfying lifestyle.[5]

Ironically, a farm evangelist could be viewed as a polar opposite to what Spillman did professionally, which was to develop the field of economics as applied to farming. "Dollar efficiency," as it was termed in his era, meant looking at the bottom line. Historian Richard S. Kirkendall calls those values "modernization" and says it ultimately replaced the agrarian tradition. It was a clash between a "romantic view" and a "more hard-headed" one. Taylorism and its gospel of efficiency pervaded most aspects of American life, particularly business and economics. Spillman's career was clearly influenced by the shifting ideology of the times. During his professional career, Spillman published more than three hundred articles, bulletins, and books, all exploring how farming could be done more efficiently with the application of scientific ideas. Yet he was always a "farm evangelist," promoting family farms and a yeoman ideal.[6]

Nevertheless, other than his genetics advances, it was his work as a farm evangelist that mattered most to those who had worked with him and come to honor his memory. His enthusiasm for agriculture and his respect for farm living had made him both respected and admired, a rare status among his contemporaries in the USDA. Rather than change people's lives completely from an expert's outsider point of view, Spillman respected the knowledge and practices farmers themselves developed. His ability to express abstract scientific ideas in a concrete form the public could understand made his work significant.

That he was able to garner respect from both the scientific community and the popular audience as well is remarkable. Few people have the ability to do both. Historian Charles Rosenberg calls such turn-of-the-century individuals "researcher-entrepreneurs" who successfully linked themselves and their work with both farm leaders and business-

5. Ibid.
6. Kirkendall, "Agricultural Colleges," 3, 14.

men, preserving the integrity of their scientific research while proving its value to the lay community. He notes that the role was so complex only a few succeeded. Rex Willard, a colleague at the Department of Agriculture and a longtime friend of Spillman, noted that it was the "versatility of his mind" that allowed him to move from a background in mathematics and physics to become a specialist in genetics and later a foremost economist. Probably more important than his wheat genetic work was his work on the law of diminishing returns. Spillman blended technical agronomic problems with animal husbandry and a dash of sociology to create a framework for envisioning agriculture. Today's term for his model would be *sustainable agriculture,* meaning a farming system that is able to maintain productivity and usefulness to society while remaining "resource-conserving, socially supportive, commercially competitive, and environmentally sound." But Spillman lived during an age driven by Frederick Winslow Taylor's efficiency theories, when the trend was to view farming as an industry focused on profitability. Spillman, however, viewed farming as a complex business system based on wise decisions that took many aspects of farming and rural living into consideration. His philosophy evolved into what he called "farm management," which was essentially a holistic view of agriculture.[7]

Spillman's ethics were unquestionably part of his success. Time and again, those who knew or worked with him mentioned his "integrity or adherence to right principles," which Willard noted. Although some records are lost regarding the various opportunities Spillman had to test those principles, what remains often hint at the problems he faced. The GEB issue loomed large in Spillman's life; his personal records contain letters of support from colleagues and newspaper clippings about that episode. Many records are no longer available because twenty cases

7. Rosenberg, *No Other Gods: On Science and American Social Thought,* 159; Willard, memorial address, October 22, 1931, Spillman Papers, MASC, 16; John Ikerd, quoted by Richard Duesterhaus in "Sustainability's Promise," 4 (available online at the Alternative Farming Systems Information Center, Sustainable Agriculture Resources, http://www.nal.usda.gov/afsic/agnic/agnic.htm [accessed December 8, 2003]); Ikerd, "The Industrialization of Agriculture: Why We Should Stop Promoting It," presented at the Harold F. Breimyer 1995 Agricultural Policy Seminar, University of Missouri, Columbia, November 16–17, 1995 (available online at http://www.ssu.missouri.edu/faculty/jikerd/papers/brsm 1–95.htm [accessed May 20, 2002]). Ikerd is professor emeritus of agricultural economics at the University of Missouri at Columbia.

of records for the Bureau of Plant Industry were destroyed while Spillman was working at the *Farm Journal.*[8]

Spillman was loyal to a fault. In fact, his loyalty to friends and stead-
fast adherence to "right principles" made his career difficult at times.
Politics was always part of the department's operation, and at the Pullman memorial service a speaker extolled Spillman for resisting the
forced resignation of fourteen men who worked under him. Spillman
fought back. "Every possible friend, every possible avenue of pressure
and attack were resorted to," Willard remembered, "with the result that
eventually the requests for the resignations were recalled." In closing,
speakers at the memorial focused on Spillman's values and principles
and their respect for the difficulties he faced at times because he refused
to compromise them. Certainly, his wheat work and advancement of
science were remarked upon, but the core of each speaker's message
was Spillman's ethics.[9]

William Spillman's work spanned more than three decades, from the
mid-1890s through the early 1930s. Agriculture underwent vast changes
in technology, markets, and landownership. During that era, U.S. agriculture had experienced a cycle of bust, boom, and again bust; had
entered into what historians call the second agricultural revolution—
the adoption of new management practices and of mechanical, biological, and chemical technologies; and the nascent appearance of farmer
political power. Spillman's work on genetics, farm management, land
tenure, and fertilizer use had contributed significantly to these devel-
opments. His domestic allotment model, as the germ of the idea for
the Agricultural Adjustment Act, helped to shape agricultural policy
from the New Deal to the present.

In 1931, Ramsay sprinkled his father's ashes on a patch of alfalfa
growing in one of the experimental plots on the Washington State College campus where he made his early scientific discoveries while working with hybrid wheats. Unlike bequeathing money to an institution,
which might have gone unnoticed as larger donations were garnered
over the years, there is something eerily memorable about choosing a

8. Willard, memorial address, 18; Asher Hobson to W. A. Taylor, August 6, 1919,
box 101, file 49, Record Group 54, Bureau of Plant Industry, "Bureau Chief's Correspondence, 1908–1939," National Archives.
    9. Willard, memorial address, 18.

public place for one's remains; his family claimed he always held a fondness for the school that gave him his start in the scientific community.

In the century since Spillman worked on the Pullman campus, agriculture and rural life have changed immensely. He certainly would lament the way private funds now dominate agricultural research and the devastating population decline in rural counties across the nation—both issues he devoted his life to resisting. Today, Mendelian crosses seem old-fashioned in the face of cell technology as biotechnology concerns divide geneticists, farmers, and consumers. At Washington State University, the agricultural classroom building erected in the 1950s on the ground that once held Spillman's experimental grass garden, and later his ashes, now houses wheat genetic laboratories. That building, Johnson Hall, is part of a major construction project, on the way to becoming a state-of-the-art biotechnology laboratory complex.

But although the new thirty-nine million–dollar facility seems to overshadow the agrarian ideals Spillman promoted, in the long run the biotechnology wave may not minimize the issues he believed in. He would be gratified to see the surge of interest in diversified, organic, family-owned farms. He would be delighted with the university-sponsored grazing programs, the Small Acreage Land Owner program offered through the Cooperative Extension Service, and the "Know Your Farmer" movement that is yielding higher prices for locally produced food across the nation. Washington State University is now a major research university, heavily invested in research to benefit industrial agriculture, but the Center for Sustaining Agricultural and Natural Resources on the campus, created by the state legislature in 1991, mirrors a movement taking place at agricultural universities across the nation. Universities and the USDA are joining the grassroots networks of farmers, consumers, and nonprofit organizations to increase support for small farms and sustainable agriculture. William Spillman would be pleased.

APPENDIX A

Publications by William J. Spillman

Most of this list was compiled by Cora Feldkamp (up to 1912), Lillian Crans (post-1912), and Ramsay Spillman (1940s), and deposited in the William J. Spillman Papers at Washington State University as well as at the National Library of Agriculture, Beltsville, Maryland. My additions to those lists are indicated with an asterisk (*).

The many articles Spillman wrote while he worked at the *Farm Journal* between 1918 and 1921 are not listed. His papers contain copies of many articles from the *Farm Journal* during this period that he likely authored, but they are not clearly attributed to him.

1886

"The Spirit of the Age." *Weekly Columbia Missouri Statesman,* March 19. This was a prize essay contest, University of Missouri.

1889

"A Comparison of the Life Histories of Different Forms of Plants" and "The Height of the Atmosphere." Presented at the annual meeting of the Indiana Academy of Sciences; no published copies found.

1890

"Geology Section at Vincennes," "Introduction of Noxious Weeds," "Preliminary List of Plants of Knox County," and "Refraction Rainbow." Presented at the annual meeting of the Indiana Academy of Sciences; no published copies found.

1892

"Above the Clouds: Perilous Ascent of Mt. Hood. Detailed Description of the Dangers and Hardships, and of the Pleasure That Awaits One at the Summit." *Portland Daily Oregonian,* September 11.

1895

The Babcock Milk Test. Bulletin 18. Pullman: Washington State Agricultural Experiment Station.
Feeding Wheat to Hogs. Bulletin 16. Pullman: Washington State Agricultural Experiment Station.
Silos and Ensilage. Bulletin 14. Pullman: Washington State Agricultural Experiment Station.

1896

The Acid Test for Milk and Cream. Bulletin 24. Pullman: Washington State Agricultural Experiment Station.
Rational Stock Feeding. Bulletin 29. Pullman: Washington State Agricultural Experiment Station.

1898

Correction of the Reading of the Babcock Test for Cream. Bulletin 32. Pullman: Washington State Agricultural Experiment Station.
Effects of Richness of Cream on Acid Test. Bulletin 32. Pullman: Washington State Agricultural Experiment Station.

1899

"The Acid Test for Milk." *American Cheesemaker* 14 (164): 4.
"The Acid Test for Milk." *Dairy and Creamery* 1 (21): 6.
*"Report of Committee on Agriculture, Oregon State Grange." In *Proceedings: 11th Annual Meeting of Washington State Grange, Mt. Pleasant,* 38–41.

1900

*Address before Washington State Grange. June 4. Subject of rural population decline. Published in *Proceedings: 12th Washington State Grange, Amboy,* 19–22.

Forage Plants in Washington. Bulletin 41. Pullman: Washington State
   Agricultural Experiment Station.
A Mechanical Ration Computer. Bulletin 48. Pullman: Washington State
   Agricultural Experiment Station.
Rational Stock Feeding. Bulletin 43. Pullman: Washington State Agri-
   cultural Experiment Station.
*"Report of the Committee on Agriculture." In *Proceedings: 12th Wash-
   ington State Grange, Amboy,* 40–41.
"The So-Called Air Churn." *Dairy Reporter* 2 (33): 526.

1902

"Exceptions to Mendel's Law." *Science* 16:709–10, 794–96.
Farmer's Institute Worker (with Discussion). Bulletin 120. Washington, DC:
   U.S. Department of Agriculture, Office of Experiment Stations.
*List of Publications of the Office of Grass and Forage Plant Investigations
   and the Division of Agrostology.* Washington, DC: U.S. Department
   of Agriculture.
*Quantitative Studies in the Transmission of Parental Characters to Hybrid
   Offspring.* Presented at meeting of Association of American Agri-
   cultural Colleges and Experiment Stations, Washington, DC,
   November 12–14, 1901. Bulletin 115. Washington, DC: U.S. Department of Agriculture, Office of Experiment Stations.
"Systems of Farm Management in the United States." In *Yearbook,* 343–
   64. Washington, DC: U.S. Department of Agriculture.

1903

Carbohydrates for Balancing the Ration for Beef Cattle in the South. Bul-
   letin 123. Washington, DC: U.S. Department of Agriculture,
   Office of Experiment Stations.
"Mendel's Law." *Popular Science Monthly* 62:269–80.
"A Model Farm." In *Yearbook,* 363–70. Washington, DC: U.S. Depart-
   ment of Agriculture.
"Quantitative Studies in the Transmission of Parental Characters to
   Hybrid Offspring." *Journal of the Royal Horticultural Society* (Lon-
   don) 27:876–93.
"Significance of the School Garden Movement." In *American Park and
   Outdoor Art Association Report,* 7:47–50.

1904

"Article on Grasses." In *Cyclopedia Americana.* New York: Munn.
"Forage Crops for Irrigated Lands." In *Proceedings: National Irrigation Congress, El Paso, Texas,* 300–303.
*"General Farming." In *Yearbook.* Washington, DC: U.S. Department of Agriculture.
"Horticultural Varieties of Common Crops." *Science* 19:34–35.
"Hybrid Wheats." *Science* 20:681.
"Mr. Detrich's Model Farm." *Hoard's Dairyman* 35:255.

1905

"A Demonstration in Methods of Farm Management." In *Proceedings: Society for Promotion of Agricultural Science,* 104–9.
Extermination of Johnson Grass. Bulletin 72. Washington, DC: U.S. Department of Agriculture, Bureau of Plant Industry.
Farm Grasses of the United States. New York: Orange Judd.
"Grasses and Forage Plants." In *Yearbook,* 587–88. Washington, DC: U.S. Department of Agriculture.
"Mendel's Law." *Medical Notes and Queries* 1 (January): 4–5.
*"Mendel's Law in Relation to Animal Breeding." In *Proceedings: American Breeders' Association,* vol. 1. Washington, DC.
"The Metric System Again." *Science* 21 (April 14): 587.
"Natural Mounds." *Science* 21 (April 21): 632.
"Opportunities in Agriculture: III, General Farming." In *Yearbook,* 181–90. Washington, DC: U.S. Department of Agriculture.
"Theoretical Studies in Breeding." In *Proceedings: American Breeders' Association,* 1:87–88. Washington, DC.

1906

"Application of Mendel's Law to a Practical Problem in Breeding Cattle." In *Proceedings: American Breeders' Association,* 2:173–77.
"Breeding the Horns Off Cattle." *Breeder's Gazette* 49 (June 20): 1287–88, 1334–35.
"Diversified Farming in the Cotton Belt." In *Yearbook,* 193–218. Washington, DC: U.S. Department of Agriculture.

An Example of Model Farming. Farmers' Bulletin 242. Washington, DC: U.S. Department of Agriculture.

"Inheritance of Color Coat in Swine." *Science* 24 (October 5): 441–43.

"Mendelian Character in Cattle." *Science* 23 (April 6): 549–51.

Renovation of Worn-Out Soils. Farmers' Bulletin 245. Washington, DC: U.S. Department of Agriculture.

A Successful Hog and Seed-Corn Farm. Farmers' Bulletin 272. Washington, DC: U.S. Department of Agriculture.

"A Valuable Bulletin on Forage Crops." *Ranch* 23 (December 15): 3.

1907

"Adjusting a Cropping System for a Maximum Herd." *Cornell Countryman* 5 (October): 10–11.

"Artificial Production of Mutants." *Science* 26 (October 11): 479.

"The Chromosome in the Transmission of Hereditary Characters." In *Proceedings: American Breeders' Association,* 3:135–37.

"Color Inheritance in Mammals." *Science* 25 (February 22): 313–14.

"The Crossing of Breeds." *Hoard's Dairyman* 38 (May 17): 427.

"Disking Alfalfa." *Hoard's Dairyman* 38 (April 12): 301–2.

"Educating Boys to Be Farmers." Address delivered at Farmers' Congress, College Station, TX, July. *Farm and Ranch* 26 (September 14): 4–5.

"The Farm Labor Question." *Hoard's Dairyman* 38 (February 22): 80.

"Grasses and Clovers in Meadows and Pastures." In *Cyclopedia of American Agriculture,* ed. L. H. Bailey, 2:442–53.

"Growing Alfalfa in the Mississippi Valley." *Hoard's Dairyman* 38 (May 24): 451.

"Impressions of New England Agriculture." *Hoard's Dairyman* 38 (March 8): 145.

"The Inheritance of the Belt in Hampshire Swine." *Science* 25 (April 5): 541, 543.

"Inheritance of the Rose Comb." *Wallace's Farmer* 32 (May 31): 680.

"Johnson Grass." In *Cyclopedia of American Agriculture,* ed. L. H. Bailey, 2:448–49.

"A Method of Adjusting the Crop Acreage to the Needs of a Herd." In *Proceedings: Society for Promotion of Agricultural Science,* 28: 139–43.

A Method of Eradicating Johnson Grass. By J. S. Cates and W. J. Spillman. Farmers' Bulletin 279. Washington, DC: U.S. Department of Agriculture.

"Monument to Mendel." *Science* 25 (March 22): 469.

"The Pennsylvania Model Farm." *Hoard's Dairyman* 38 (May 24): 451.

Planning a Cropping System. Bulletin 102. Washington, DC: U.S. Department of Agriculture, Bureau of Plant Industry.

"Possibility of the Agricultural High School." *Southern Highlander* 1 (April): 3.

"Potatoes on Dairy Farms." *Hoard's Dairyman* 38 (August 30): 773–83.

"Report of Committee on Animal Hybridizing." In *Proceedings: American Breeders' Association,* 3:184–91.

"A Sheep Goat Hybrid." *Science* 25:791–92.

"Standardizing Breed Characteristics." In *Proceedings: Society for Promotion of Agricultural Science,* 28:115–21.

A Successful Alabama Diversification Farm. By M. A. Crosby, J. F. Duggar, and W. J. Spillman. Farmers' Bulletin 310. Washington, DC: U.S. Department of Agriculture.

"Types of Farming and Their Possibilities." Address given at annual meeting of Indiana Corn Growers Association, December 16. *Prairie Farmer* 79 (December 19): 1010–11.

1908

"Agronomic Habits of Root Stock Producing Weeds." By W. J. Spillman and J. S. Cates. In *Proceedings: Society for the Promotion of Agricultural Science,* 29:57–66.

"Biotypes of Corn." *Science* 28 (July 17): 88.

"Color Factors in Mammals." In *Proceedings: American Breeders' Association,* 4:357–59.

"Cropping Systems for Stock Farms." In *Yearbook,* 385–98. Washington, DC: U.S. Department of Agriculture.

"Crop Rotation." *Southern Planter* 69 (August): 688–89.

"Crop Rotation in Illinois." *Hoard's Dairyman* 39 (July 10): 659.

"Farm Management." *Farm and Fireside* 31 (August 25): 1–2.

"How to Keep the Boys on the Farm." In *Proceedings: Texas Farmers' Congress,* 10:23–26.

"Inheritance of Fluctuating Variations." *Science* 27 (March 27): 509–10.

"An Interpretation of Elementary Species." *Science* 27 (June 5): 896–98.

"Mendelian Phenomena and Discontinuous Variation." In *Proceedings: American Breeders' Association,* 4:359.

"Mendel's Law." *Wallace's Farmer* 33 (May 8): 654.

"Notes and Literature: Heredity." *American Naturalist* 42 (September).

"Order of Agricultural Development." *Horn and Hoof* 1 (July): 1–3.

"Origin of Varieties of Domesticated Species." *Science* 28 (August 21): 252–54.

"Pedagogical Value of Farm Work." *Southern Highlander* (March): 42–44.

"Present Condition of the Detrich Farm." *Hoard's Dairyman* 39 (October 2): 957.

"Report of the Committee on Animal Hybridizing." In *Proceedings: American Breeders' Association,* 4:317–23.

"Some Dairy Farm Problems." *Vermont Dairymen's Association* 38:99–96.

"Sowing Clover in Corn and Keeping Garlic Flavor Out of Milk." *Hoard's Dairyman* 29 (May 15): 447–49.

"Spurious Allelomorphism (in Poultry): Results of Some Recent Investigation." *American Naturalist* 42 (September): 610–15.

"Teaching Agriculture in Public Schools." *Wallace's Farmer* 33 (December 25): 1601.

1909

Application of Some of the Principles of Heredity to Plant Breeding. Bulletin 165. Washington, DC: U.S. Department of Agriculture, Bureau of Plant Industry.

"Barring in Plymouth Rocks." *Poultry* 6 (October): 7–8, 14.

"The Country Boy." *Science* 30 (September 24): 405–7.

"The Country Boy Again." *Science* 29 (May 7): 739–41.

"Education and the Trades." *Science* 29 (February 12): 255–56.

"The Effect of Different Methods of Selection on the Fixation of Hybrids." In *Proceedings: American Breeders' Association,* 5:341–47.

"Heredity: A Case of Non-Mendelian Heredity." *American Naturalist* 43 (July): 437–48.

"The History of the Mule-Footed Hog." *Science* 30 (December 10): 855–56.

The Hybrid Wheats. Bulletin 89. Pullman: Washington State Agricultural Experiment Station.

"International Language." *Science* 30 (December 10): 841–43.

"Is the Present System of Tenant Farming Building a Strong System of Agriculture?" *Hoard's Dairyman* 40 (July 9): 698–99.

"Methods of Killing Quack Grass." *Hoard's Dairyman* 40 (February 26): 124.

"The Nature of 'Unit' Characters." *American Naturalist* 43 (February 12): 243–48.

"New England Dairy Farming." *Hoard's Dairyman* 40 (April 9): 345.

"Notes and Literature: Heredity." *American Naturalist* 43 (April).

"Origin of Polled Durhams." *Breeder's Gazette* 55 (March 31): 778.

"Progress of the International Language Esperanto." *Science* 30 (October 8): 478–79.

"Recent Advancement in Our Knowledge of the Laws of Heredity." Presented at meeting of American Breeders' Association, Columbia, MO, January 6–8. In *Proceedings: American Breeders' Association,* 5:78–93.

A Successful Poultry and Dairy Farm. Farmers' Bulletin 355. Washington, DC: U.S. Department of Agriculture.

"Types of Farming in the United States." In *Yearbook,* 351–66. Washington, DC: U.S. Department of Agriculture.

1910

"Back to the Land." *Farm World* 4 (9): 13; (10): 2; (12): 19; 5 (1): 19; (2): 11; (3): 14; 6 (1): 17; (8): 13.

"Basis for Estimating the Yield of Hay." In *Proceedings: American Society for Agronomy,* 1:158–59.

"The City Man as a Farmer." *Leslie's Weekly* (December 15): 635.

"Cost Accounting in Dairy Farming." *Hoard's Dairyman* 41 (April 8): 376–78.

"The Esperanto Congress." *Scientific American* 103 (September 3): 179.

"The Essentials in the Training of the Investigator." By A. C. True, H. P. Armsby, W. H. Jordan, F. E. Thorne, and W. J. Spillman. In *Proceedings: Society for Promotion of Agricultural Science,* 31:122–23.

"Farm Bookkeeping." Parts 1–2. *National Stockman and Farmer* 34 (December 15): 960–61; (December 22): 991.

"Farming as an Occupation for City-Bred Men." In *Yearbook,* 239–49. Washington, DC: U.S. Department of Agriculture.

"Good Farming and Attractive Country Homes." *Twice-a-Week Spokesman-Review.* Comp. Edwin Augustus Smith. Contains the following by W. J. Spillman: "Help to Solve Farm Problems," 8–9; "An Ideal Farm House," 10–12; "Ideal of Ten-Acre Irrigated Farm," 18–24; "Discussion of Prize Farm House Plans," 121–24.

"Heredity." *American Naturalist* 44 (August): 504–12.

"History and Peculiarities of the Mule-Foot Hog." *American Breeders' Magazine* 1:178–82.

"Is Dairy Farming Profitable?" *Hoard's Dairyman* 41 (April 15): 404–5.

"Mendelian Phenomena without De Vriesian Theory." *American Naturalist* 44 (April): 214–28.

"The Mendelian View of Melanin Formation." *American Naturalist* 44 (February): 116, 123.

"Notes on Heredity and Evolution." *American Naturalist* 44 (December): 750–62.

"Report of Committee on Animal Hybrids." *American Breeders' Magazine* 1:193–96.

"Report of the Evolution Committee of the Royal Society (London)." *American Naturalist* 44 (August): 504–12.

"A Rotation of Quack Grass Land." *Hoard's Dairyman* 41 (May 13): 509.

"Selection of Vegetatively Propagated Crops." In *Proceedings: American Society for Agronomy,* 1:90–94.

"Shall the Dairyman Buy Concentrate?" Parts 1–2. *Hoard's Dairyman* 41 (February 25): 123–24; (April 8): 370–78.

"The Soil as a Limiting Factor in Crop Production." In *Proceedings: American Society for Agronomy,* 1:211–17.

Soil Conservation. Farmers' Bulletin 406. Washington, DC: U.S. Department of Agriculture.

"A Theory of Mendelian Phenomena." *American Breeders' Magazine* 1:113–25.

1911

"Application of the Principles of Heredity to the Improvement of Plants and Animals." In *Proceedings: American Breeders' Association,* 6:397–419.

"Back to the Land." *Farm World* 4 (9): 13; (10): 2; (12): 19; 5 (1): 19; (2): 11; (3): 14; 6 (1): 17; (8): 13.

"Cost of the Farm Dwelling." *Hoard's Dairyman* 42 (August 11): 858–59.

"Creating New Animals and Plants." *Scientific American* 104 (February 18): 164–65, 184–85.

"Efficient Farm Management: The Newest Methods for Increasing Profits." *Country Gentleman* 76 (November 25): 10.

"History and Peculiarities of the Mule-Foot Hog." *California Cultivator* 36 (January 5): 12.

"History and Peculiarities of the Mule-Foot Hog." In *Proceedings: American Breeders' Association*, 6:116–20.

"An Important Principle in Selecting for Fancy Points." In *Proceedings: American Breeders' Association*, 6:375–80.

"Improvement of Plants and Animals by Breeding and Selection." In *New Jersey State Board of Agriculture Report*, 38:151–53.

"Inheritance of the 'Eye' in *Vigna*." *American Naturalist* 45 (September): 513–23.

"Notes on Heredity." Parts 1–2. *American Naturalist* 45 (January): 60–64; (August): 507–12.

"Organising the Farm for Profit." *New York Tribune Farmer* 11 (November 23): 1, 11.

"Reformed Calendar." *Science* 33 (June 2): 854.

"Report of Committee on Animal Hybrids." In *Proceedings: American Breeders' Association*, 6:131–34.

"Some Present Agricultural Problems." In *Kansas State Board of Agriculture Biennial Report*, 17:3–9.

"A Theory of Mendelian Phenomena." In *Proceedings: American Breeders' Association*, 6:78–90.

1912

"Chromosomes in Wheat and Rye." *Science* 35 (January 19): 104.

"Competition between Alfalfa and Corn." *Hoard's Dairyman* 43 (July 26): 894.

"Determining the Mean Length of Life: A Method Applicable to Living Population and to Commodities Having a Limited Period of Usefulness." *Scientific American*, supp., 74 (December 14): 378.

"The Difficulty of Changing Agricultural Practice." *Cornell Countryman* 9 (January): 99–101.

"Emmer for the South." *Country Gentleman* 77 (April 20): 17.

"Farm Management Experts for Each County: Government Plans to Place an Expert at the Service of Farmers in Agricultural Counties Which Are Willing to Cooperate." *New York Tribune Farmer* 11 (June 13): 4.

"Farm Organization: A Discussion of the Principles Involved in Farm Organization." *Wallace's Farmer* 37 (October 18): 1480.

"The Farm Problem Extension Work of the United States Department of Agriculture." In *Proceedings: Association of American Agricultural Colleges and Experiment Stations*, 25:95–97.

"Heredity." Parts 1–3. *American Naturalist* 46 (February): 110–20; (March): 163–65; (May): 309–12.

"A Method of Determining the Average Length of Life of Farm Equipment." *Science* 36 (October 25): 565–68.

"A New Plan to Unit Extension Work: How to Avoid Conflict between State and National Agencies." *Country Gentleman* 77 (September 21): 5.

"The Organization and Redirection of American Agriculture." *Continental* 1 (April): 21–26.

"The Present Status of the Genetics Problem." *Science* 35 (May 17): 757–67.

"Seasonal Distribution of Labor on the Farm." Abstract. Parts 1–3. *New York Tribune Farmer* 11 (June 27): 2; (July 4): 4; (July 11): 4.

"Seasonal Distribution of Labor on the Farm." In *Yearbook*, 269–84. Washington, DC: U.S. Department of Agriculture.

"Some Experiences on a Missouri Hog Farm." *Southern Planter* 73 (September): 962–63.

"Study of Farm Practice versus Field Experiments." In *Proceedings: Society for Promotion of Agricultural Science*, 33:103–13.

What Is Farm Management? Bulletin 259. Washington, DC: U.S. Department of Agriculture, Bureau of Plant Industry.

"What Is Farm Management? How Does It Help the Practical Man to Solve His Problem?" *New York Tribune Farmer* 11 (May 23): 3, 17.

1913

"Are Land Values Too High?" *Breeder's Gazette* 64 (December): 1205, 1280.

"Color Correlation in Cowpeas." *Science* 38 (August 29): 302.

"Conservation of Rainfall: Memorandum on the Work of Col. Freeman Thorp on His Farm at Hubert, Minnesota." *Congressional Record,* 63d Cong., 1st sess: S 228.

The Farmer's Income. Bulletin 132. Washington, DC: U.S. Department of Agriculture, Bureau of Plant Industry.

"Farm Management in the United States at the Present Time: The Actual Scope of Its Work, and Recent Development Therein." *International Institute of Agriculture Bulletin: Agricultural Institute and Plant Diseases* 4 (July): 997–1000.

"The Fundamental Problems of American Farm Management." *Business America* 13 (February): 154–58.

Future Meat Supply of the United States. Farmers' Bulletin 560. Washington, DC: U.S. Department of Agriculture.

"Government Hay Measuring Rules." *Breeder's Gazette* 64 (August 20): 310–11.

"Inheritance of the Poll Character in Cattle." *Jersey Bulletin* 32 (January 29): 168–71.

"The Law of Recombination." *Breeder's Gazette* 64 (October 23): 776.

"Marketing Farm Products." *Creamery and Milk Plant Monthly* 1 (May): 16–18.

"Marketing Farm Products." In *Proceedings: 1st National Conference on Marketing and Farm Credits,* 109–21.

"Measuring Hay in Ricks or Stacks." By H. B. McClure, W. J. Spillman, and J. W. Froley. Circular 131. Washington, DC: U.S. Department of Agriculture, Bureau of Plant Industry.

"Mendel's Laws." *American Agriculturist* (December 12).

"Pasture Following Rye." *Hoard's Dairyman* 45 (February 28): 186.

"The Place of Small Grain on the Farm and the Management of Crops of This Character." *Ohio Farmer* 131 (February 15): 209–10.

"The Problem of Tenant Farming." *Southern Planter* 74 (December): 1197–98.

"The Relation of Farm Management Work to Other Departments of the College and Experiment Station." In *Proceedings: Association of American Agricultural Colleges and Experiment Stations,* 26: 166–72.

"Review of Ridgway's Color Standards and Nomenclature." *Science* 37 (June 27): 985–89.

"Sex-Limited Inheritance." *Breeder's Gazette* 64 (October 30): 824.

"What Is the Relative Effectiveness of Different Means of Reaching the Farmer?" In pamphlet *What the U.S. Department of Agriculture Wanted to Know Turned Out to Be What You Want to Know.* New York: Farm and Fireside.

1914

"Building a Farm while Making a Profit." *Southern Planter* 75 (March): 201–3.

"Cost of Growing Apples." In *Proceedings: New York State Fruit Growers' Association,* 13:115–19.

"Cost of Growing Apples." *National Stockman and Farmer* 37 (February 21): 18.

"Cost of Growing Cotton." *Farm Management Monthly* 2 (November): 115.

"Data to Be Gathered in Farm Management Surveys." In *Proceedings: 4th Annual Meeting American Farm Management Association,* 49.

"Effect of the Smith-Lever Bill on the Working of the U.S. Department of Agriculture." *Quarterly of Alpha Zeta* 13 (December): 9–10.

"Factors of Efficiency in Farming." In *Yearbook,* 93–108. Washington, DC: U.S. Department of Agriculture.

"The Farm Score Card." *U.S. Department of Agriculture: Weekly Newsletter to Crop Correspondents* 2 (August 12): 3–4.

"Future Meat Supply of the United States." *Missouri and Kansas Farmer* 410 (October 1): 428.

"Future Meat Supply of the United States." *Northwest Pacific Farmer* 45 (October 22): 1.

"Increases in Farm Land Values." *Breeder's Gazette* (June 4).

"Length of German Working Days." *Breeder's Gazette* 65 (April 16): 873.

"Mendel Laws of Breeding Explained." *Southern Farming* 94 (August 1): 2–3.

"The Silo as an Investment." In *Nebraska Corn Improvers' Association 5th Annual Report,* 84–89.

"Types of Farming in Chester County, Pennsylvania, and Western Central Illinois." *Farm Management Monthly* 2 (December): 120–21.

1915

"The Combined Harvester and Thresher." *U.S. Department of Agriculture: Weekly News Letter* 2 (July 28): 1.

"The Efficiency Movement in Its Relation to Agriculture." *American Academy of Political and Social Science Annals* 59 (May): 65–76.

"The Farmers Response to Economic Forces." *Economic World* 10:362–65. Also in *Proceedings: 6th Annual Meeting, American Farm Management Association,* 12–14.

"Farm Organization Investigations and Their Relation to the Farm Survey." In *Proceedings: 5th Annual Meeting, American Farm Management Association,* 12–14.

"Formulae for Calculating Interest on Farm Equipment." Circular 53. Washington, DC: U.S. Department of Agriculture, Office of the Secretary.

"Fundamental Factors in Rural Improvement." *University of California Journal of Agriculture* 2 (November): 94–95.

"How the Farmer Can Use the Facilities of the U.S. Department of Agriculture." *U.S. Department of Agriculture: Weekly News Letter* 2 (January 13): 2–3.

"A Method of Calculating the Percentage of Recessives from Incomplete Data." *American Naturalist* 49 (June): 383–84.

"The Office of Farm Management." *Quarterly of Alpha Zeta* 13 (March): 80–81.

"A Perpetual Calendar." *Scientific American,* supp., 79 (February 27): 141.

"Small Farm." *Farm Management Monthly* 3 (February): 131.

A Theory of Gravitation and Related Phenomena. Lancaster, PA: New Era Printing.

1916

"The Average Internal Curve and Its Application to Meteorological Phenomena." By W. J. Spillman, H. R. Tolley, and W. G. Read. *Monthly Weather Review* (U.S. Department of Agriculture Weather Bureau) 44 (April): 197–200.

*"The Farmer's Income." In *Selected Readings in Rural Economics,* comp. T. N. Carver, 630–35. New York: Ginn.

"Farm Management and National Efficiency." In *Proceedings: 7th Annual Meeting, American Farm Management Association,* 56–59.

Farm Management Practice of Chester County, Pennsylvania. By W. J. Spillman, H. M. Dixon, and G. A. Billings. Bulletin 341. Washington, DC: U.S. Department of Agriculture.

"Getting Along on Small Places." *Field and Farm* 31 (January 8): 4.

"Measuring Hay in Ricks or Stacks." By H. B. McClure and W. J. Spillman. Circular 67. Washington, DC: U.S. Department of Agriculture, Office of the Secretary.

"Report of the Committee on Wheat and Corn." In *Proceedings: National Conservation Congress.* Washington, DC.

Suggestions Concerning Checking and Tabulating Farm Management Survey Data. Circular 1. Washington, DC: U.S. Department of Agriculture, Office of Farm Management.

1917

"Biology and the Nation's Food." *Scientific Monthly* 4 (March): 220–25.

"Farm Tenantry in the United States." By W. J. Spillman and E. A. Goldenweiser. In *Yearbook,* 321–46. Washington, DC: U.S. Department of Agriculture.

Human Food from an Acre of Staple Farm Products. By M. O. Cooper and W. J. Spillman. Farmers' Bulletin 877. Washington, DC: U.S. Department of Agriculture.

"Measuring Hay." *Wallace's Farmer* 42 (February 23): 346.

Plan for Handling the Farm-Labor Problem. Circular 2. Washington, DC: U.S. Department of Agriculture, Office of Farm Management.

"Potatoes on Dairy Farms." *Hoard's Dairyman* 38 (August 30): 773–83.

Validity of the Survey Method of Research. Bulletin 529. Washington, DC: U.S. Department of Agriculture.

"What Would You Do?" *Country Gentleman* 82 (May 12): 4.

1918

Factors of Successful Farming Near Monett, Missouri. Bulletin 633. Washington, DC: U.S. Department of Agriculture.

Farm Diary: A Business Record and Account Book. By E. H. Thomson, with introduction by W. J. Spillman. Yonkers-on-Hudson, NY: World Book.

"Farm Management Aspects of the Potato Crop." *Potato Magazine* 1 (June): 9.

Farm Science: A Foundation Textbook on Agriculture. Yonkers-on-Hudson, NY: World Book.

"How Farmers Acquire Their Farms." In *Proceedings: Society for Promotion of Agricultural Science,* 38:87–90.

Measuring Hay in Ricks or Stacks. By H. B. McClure and W. J. Spillman. Bulletin 23. Moscow: University of Idaho Extension Service.

"A New Calendar." *Science* 47 (May 17): 488–89.

*"Sick Farms Made Well." *Farm Journal* (November): 5.

*"Sick Farms Made Well." *Farm Journal* (December): 8.

"Work of the Office of Farm Management Relating to Land Classification and Land Tenure." *American Economic Review* 8 (1): 65–71. Also in *Proceedings: 8th Annual Meeting, American Farm Management Association,* 40–46. Also in *Proceedings: 30th Annual Meeting of the American Economic Association,* 65–71.

1919

"The Agricultural Ladder." *American Economic Review* 9:1, supp., *Papers and Proceedings of the Thirty-first Annual Meeting of the American Economic Association* (March): 170–79. Also in *Federal Board for Vocational Education: Vocational Summary* 1 (9): 19–21.

"Dr. W. J. Spillman Tells Where to Find the Best Farm Knowledge." *Oregon Farmer* 29 (June 18): 3, 6.

"Farm Management—a New Science." *Tractor Farming* 4 (April): 6, 12–13. Also in *Utah Farmer* 15 (July 5): 818–19. Reprinted in *Essays on Agriculture,* by S. D. Babbitt and L. C. Wimberly, 187–94.

"Just Settlement of the War between Producers and Consumers." *Columbia University Quarterly* 21 (April): 98–109.

"Land Tenure and Public Policy—Discussion." By W. J. Spillman, Charles L. Stewart, and B. H. Hibbard. In *Papers on Tenancy,* Office of the Secretary of the American Association for Agricultural Legislation, University of Wisconsin, Madison. Bulletin 2 (March): 72–77. Also in *American Economic Review* 9 (March): 226–31.

*"Sick Farms Made Well: Taking the Temperature of a Sick Farm." *Farm Journal* (January): 8.

*"Sick Farms Made Well: Curing a Farm of the Cramps." *Farm Journal* (February): 5.

*"Sick Farms Made Well: The Farm with Consumption." *Farm Journal* (March): 8.

*"Sick Farms Made Well: Curing a Farm of Anemia." *Farm Journal* (May): 10.

*"Sick Farms Made Well: Curing a Farm of Cancer." *Farm Journal* (June): 6.

*"Sick Farms Made Well: Curing a Farm of Rheumatism." *Farm Journal* (July): 6.

*"Sick Farms Made Well: The Farm with Appendicitis." *Farm Journal* (August): 6.

*"Sick Farms Made Well: A Farm Business Struck by Apoplexy." *Farm Journal* (September): 6.

*"Sick Farms Made Well: Curing a Farm of Pinkeye." *Farm Journal* (October): 7.

*"Sick Farms Made Well: Curing a Farm of Malaria." *Farm Journal* (November): 12.

*"Sick Farms Made Well: Burying a Dead Farm." *Farm Journal* (December): 12.

Why Data on Cost of Production Are "Not Reliable." 65th Cong., 3rd sess. (February 25): H Res 611.

1920

"Formulae Giving the Day of the Week of Any Date." *Science* 51 (May 21): 513–14.

*"Gullies, Goats and Sticky Soil." *Farm Journal* (May): 66.

*"National Unrest and the Remedy." *Farm Journal* (January): 7.

1921

"Farmer Control of Farm Prices." *Missouri State Board of Agriculture: Monthly Bulletin* (Jefferson City) 19 (April).

"A Plan for the Conduct of Fertilizer Experiments." *Journal of American Society of Agronomy* 13 (November): 304–10.

1922

"Discussion on Development of American Farm Economic Association." *Journal of Farm Economics* 4 (April): 99–100.

A Picture of the Grain Industry: Crop Areas—Buying Areas—Future of the Industry . . . Marketing Grain. Chicago: American Institute of Agriculture. "Confidential edition issued for members."

1923

"Application of the Law of Diminishing Returns to Some Fertilizer and Feed Data." *Journal of Farm Economics* 5 (January): 36–52.

Distribution of Types of Farming in the United States. Farmers' Bulletin 1289. Washington, DC: U.S. Department of Agriculture.

"Oats, Barley, Rye, Rice, Grain Sorghums, Seed Flax and Buckwheat." By C. R. Ball, T. R. Stanton, H. V. Harlan, C. E. Leighty, C. E. Chambliss, A. C. Dillman, O. C. Stine, O. E. Baker, O. A. Juve, and W. J. Spillman. In *Yearbook,* 469–568. Washington, DC: U.S. Department of Agriculture.

1924

"Application of the Law of Diminishing Returns to Some Fertilizer Data." *American Fertilizer* 60:42, 44–46, 51–54.

"Artichoke Sugar: A Discovery That Beet Folks Should Watch." *Country Gentleman* 89 (December 27): 14, 28.

Balanced Farming for the Inland Empire: The Story of Dr. Spillman's Lectures during the Agricultural Tour of 1924. Spokane: Chamber of Commerce, Agricultural Bureau. (Does not contain the actual lectures, but summaries of the lectures by Fred W. Clemens, Spokane).

"Corn Is King More Than Ever." *Farm Journal* (February).

Farm Management. New York: Orange Judd.

"Farm Ownership and Tenancy." By L. C. Gray, C. L. Stewart, H. A. Turner, J. T. Saunders, and W. J. Spillman. In *Yearbook,* 507–600. Washington, DC: U.S. Department of Agriculture.

"The Flax Grower and the Building Boom." *Country Gentleman* (August 30).

"The Golden Fleece: Sheepmen Should Watch the Clothing Trade and the Artificial-Silk Industry." *Country Gentleman* 89 (44): 7, 20.

"How Much Fertilizer?" *Potato News Bulletin* (July).

"Humpty Dumpty Had a Fall: Maybe a Shift Is Coming in Egg Farming." *Country Gentleman* 89 (30): 8, 28.

The Law of Diminishing Returns. By W. J. Spillman and Emil Lang. Yonkers-on-Hudson, NY: World Book, 1924.

"Law of the Diminishing Increment in the Fattening of Steers and Hogs." *Journal of Farm Economics* 6 (2): 166–78.

"A New Dual-Purpose Idea." *Breeder's Gazette* (March 13).

"No Corn to Burn." *Country Gentleman* (November 22).

The Problem of Diversified Crops in the West: Its Limitations and Possibilities. Washington, DC: U.S. Department of Agriculture, Bureau of Agricultural Economics.

"Wheat without Risk." *Farm Journal* 48 (October): 9, 92.

1925

"Applications of the Law of Diminishing Returns to Agronomic Problems." *Journal of American Society of Agronomy* 17 (April): 189–98.

"A Balanced Agricultural Output in the United States." *American Academy of Political and Social Science Annals* 117 (January): 285–92.

"The Boom in Beef Cattle: Next Season Promises Better Things for Both Feeders and Range Men." *Country Gentleman* 90 (February 14): 6, 51.

"Combing the Earth for Cotton: High Prices Fix Foreign Eyes on the Crop." *Country Gentleman* 90 (June 13): 5, 37.

"The Core of the Apple: Growers Can Do Much to Fit the Size of the Crop to the Profitable Demand." *Country Gentleman* 90 (May 23): 8, 34.

"Corn, Hogs, and the Price of Butter." *Country Gentleman* 90 (April 25): 10, 32.

"Hay." By C. V. Piper, R. A. Oakley, H. N. Vinall, A. J. Pieters, W. J. Morse, W. J. Spillman, O. C. Stine, J. S. Cotton, G. A. Collier, M. R. Cooper, E. C. Parker, E. W. Sheets, and A. T. Semple. In *Yearbook*, 285–376. Washington, DC: U.S. Department of Agriculture.

"The Potato Seesaw: This Is a Year for Cautious Planting." *Country Gentleman* 90 (March 7): 9, 28.

"Raw Cotton Resources." *Harvard Business Review* 3 (July): 466–73.

1926

Farming in the Big Bend Country. Popular Bulletin 135. Pullman: Washington State Agricultural Experiment Station.
"Influence of Air and Sunshine on the Growth of Trees." *Science* 63 (January 1): 18.
"Next Year's Potatoes." *Country Gentleman* 91 (February): 84.
"Study of Maladjustments in Specific Areas." *Journal of Farm Economics* 8 (January): 118–25.

1927

Balancing the Farm Output: A Statement of the Present Deplorable Conditions of Farming, Its Causes and Suggested Remedies. New York: Orange Judd.
"Potatoes in 1927." *Country Gentleman* 92 (February): 102.
"Type of Agricultural Production Affecting Expenditures and Culture." In *Farm Income and Farm Life,* by the American Country Life Association. Chicago: University of Chicago Press.
"What Is a Satisfactory Standard of Living for the Farmer?" In *Farm Income and Farm Life,* by the American Country Life Association. Chicago: University of Chicago Press.

1928

"Acreage of Cotton Moves West." *Oklahoma Farmer–Stockman* 41 (March 1): 6, 29.
Changes in Southern Agriculture and the Problems Arising Therefrom. Address before the Economics Section, Association of Southern Agricultural Workers, Memphis, February 2. Mimeograph copy. Washington, DC: U.S. Department of Agriculture, Bureau of Agricultural Economics.
*"Extra-Dry Farming." *Farm Journal* (March): 22.
The Problem of Indian Administration: Summary of Findings and Recommendations from the Report of a Survey Made at the Request of Hubert Work, Secretary of the Interior and Submitted to Him Feb. 12, 1928. By Director Lewis Meriam, W. J. Spillman, and others. Washington, DC: Brookings Institution.

Type-of-Farming Studies. By W. J. Spillman and E. E. Elliott. Mimeograph copy. Washington, DC: U.S. Department of Agriculture, Bureau of Agricultural Economics, Division of Farm Management and Costs.

1929

Balancing the Livestock Outfit. Radio talk. Mimeograph copy. Washington, DC: U.S. Department of Agriculture, Bureau of Agricultural Economics, Division of Farm Management and Costs.

"Beating the Surplus in Butter and Eggs." *Country Gentleman* 94 (December): 19, 88.

"How Much Fertilizer for Profit?" *Farm and Ranch* 48 (December 14): 9.

Locating Counties in a State. Washington, DC: U.S. Department of Agriculture, Bureau of Agricultural Economics, Division of Farm Management and Costs.

"Means of Preventing Surplus Production." In *Proceedings: Virginia Agricultural and Mechanical College and Polytechnic Institute, Institute of Rural Affairs,* 25–39. Also issued in mimeograph form. Washington, DC: U.S. Department of Agriculture, Bureau of Agricultural Economics, Division of Farm Management and Costs.

"Placing the Farm on Wheels." *Southern Planter* 90 (November 1): 4, 18–21.

"The Present Economic Status and Future Prospects of the Fruit and Vegetable Industry." In *Proceedings: New Jersey State Horticultural Society,* 93–101.

"Recent Trends Balancing Agriculture in the United States." *American Academy of Political and Social Science Annals* 142 (March): 210–15.

"The Shift in Cotton Acreage." *Farm and Ranch* 48 (November 2): 3, 11.

1930

"Measuring Absorbed Phosphates and Nitrogen." *Science* 73:215–16.

"Mendelian Factors in the Cowpea (*Vigna* species)." By W. J. Spillman and William J. Sando. Reprint from papers of *Michigan Academy of Science, Arts and Letters,* 9:249–83.

"The Need of Reorganization in Agriculture." *Agricultural Engineering* 12 (1): 17–19. Presented at meeting of Power and Machinery Division of the American Society of Agricultural Engineers, Chicago, December.

"A New Basis for Fertilizer Experiments." *Science* 71:135–36.

Shifts in Farming in the United States: A Preliminary Report. Mimeograph copy. Washington, DC: U.S. Department of Agriculture, Bureau of Agricultural Economics, Division of Farm Management and Costs.

"Types of Farming in the United States." In *Proceedings: 2nd International Conference of Agricultural Economists,* 807–12. Menasha, WI: Collegiate Press.

1933 (POSTHUMOUS)

Use of the Exponential Yield Curve in Fertilizer Experiments. Technical Bulletin 348. Washington, DC: U.S. Department of Agriculture.

APPENDIX B

"Emergency Bulletin"

Following is a typescript of the pseudoscientific bulletin prepared by William Spillman's colleagues in the Office of Farm Management when he quit as chief in 1918. It is a spoof on both his work at the department as well as the sorts of studies being done at the time. Spillman returned to the USDA later, as a consultant under Secretary Henry C. Wallace. Copies are in the William J. Spillman Papers at Manuscript, Archives, and Special Collections, Washington State University, as well as the Special Collections at the National Agriculture Library, Beltsville, Maryland.

* * *

Farm Management Investigations as Applied to the Measurement of Terrestrial and Sidereal Time in Accordance with the Principles of Economics and Psychology.

by
The Staff of the Office
-EMERGENCY BULLETIN-

Office of the Secretary
Contribution from the Office of Farm Management
W. J. Spillman, Chief.

Washington, D.C. June, 1918.

This bulletin is compiled from the unwritten remarks and unspoken thoughts of the Office of Farm Management field men in moments of joy and gloom. It was officially read before Professor Spillman on the occasion of his regrettable yielding to the siren of agricultural journalism August 31, 1918. The manuscript shows something of the scintillation that may be expected from the Office of Farm Management workers when they are relieved of the incubus of a publication committee. This manuscript has not been censored. Hence the springiness, life, personal touch, intimate revelations and bold treatment.

Contents

Approved by
[W. J. Spillman]
Chief.

Farm Practice in the Use and Care of the Ingersoll Watch

Since the Ingersoll has become a part of the regular farm organization the necessity for first hand information concerning the most approved farm practice in the use and care of this important equipment becomes apparent.

The Ingersoll is now used for much the same purpose as other watches except for telling the time. It will wear out clothes, drive nails and amuse the baby, but many of the best farmers find that it can not be relied on to tell time except when the sun is on the meridian.

In order to avoid accidents and the loss of valuable time the following simple precautions shall be observed by all beginners. Don't expect too much. You may be disappointed. The watch has no self-starter.

Therefore, before attempting to start it see that all the bearings are tight, and have the mechanism well lubricated with unsalted butter. Grasp the mud guard firmly in the left hand, and turn the crank briskly with the right for fifteen minutes. If this treatment does not give results, consult the instructor's book and repeat the operation at regular intervals through the day. The first few revolutions of the fly wheel may be accompanied by an unusual noise and much smoke. In case it backfires and lacks power, apply a wet blanket and telephone the nearest service station, or notify the Office of Farm Management.

Farm practice in the use of the Ingersoll varies widely with kind of soil, altitude, rainfall and temperature of the operator. Moorhouse found that 90% of the sugar beet raisers of Utah carried the watch in a pants pocket, while 9% left it in the kitchen and 1% wore it on the wrist or ankle. On this group of farms mica, axle grease and oleomargarine ran neck and neck as lubricants. In starting the watch it may be knocked against a hoe-handle or forced under pressure of the auxiliary engine. Miller from personal experience states that in winding there is real danger unless one takes precaution against a possible kick back.

The Ingersoll has other uses than as a farm machine. Dr. Wiley's babies used an Ingersoll as a tooth cutter. Thousands of German bullets and shells have already rebounded from Ingersoll watches in the pockets of our soldiers.

Funk's recent survey of the beer and truck gardens revolving around Cincinnati showed that no native born American was ever annoyed by a book agent while carrying an Ingersoll.

The Cost of Time by the Ingersoll Method

In a recent survey of 256 owner operators located in a representative area of Georgetown it was found that the average cost of time by the Ingersoll method is $164.53 per year.

The items that go to make up this cost may be grouped under four main headings.

> Operating Expenses
> Repairs
> Interest and depreciation
> Insurance & Taxes

Operating Expenses

Fuel and Oil—
    Gasoline . . . . $ .52
    Unslaked lime 06
    Mica axle grease 21.00

Labor—
    Winding watch (180 man hours at 40 cents/hour) 72.00
    Setting watch (4 times daily at 2 cents/set) 29.20
    Asking others the correct time (15 times daily at
    1/2 cent per time) 27.78
                                               150.56

Repairs—
    One main spring 12
    One counter balanced steering wheel 7.00
    One safety valve 10
    Full set of differential pinions 30
    One pair of skid chains 3.00
    Two blowout patches 45
                                                 10.97

Interest and Depreciation—
    Depreciation in one year 100% or 2.00
    Interest on investment of 6% 12
                                               2.12

Insurance and Taxes—
        Insurance against theft, breakage, fire, explosion,
        desertion, rust, moths, lightning, cyclone, bail and
        souvenir hunters . . . . $ .26
    Regular property tax 02
    Interstate license 60
    War revenue tax 10

Special Ingersoll surtax 09
Permit for carrying Ingersoll on Pullman cars 04

$\overline{}$

.91

Credits—
Credits to be deducted are few and uncertain—
Appreciation in value . . . .   none
Wool and hides . . . .   "
Manure . . . .   "
Pawn shop value . . . .   "
Ferrous oxide (iron rust) 02
Dirt and old grease 01

These results are similar to the figures obtained by Yerkes on the cost of operating a tractor in Illinois.

Moorhouse in Colorado finds that it costs less to grow an acre of sugar beets than it does to operate an Ingersoll.

The tabulations made by A. G. Smith on the Anderson County survey clearly show that the family mule can tell when it is dinner time without looking at any watch at all.

Formula for Calculating Ingersoll Time

As an example of the striking applicability to practical affairs of farm management investigations even of the most abstruse nature we have in the following mathematical calculation merely transformed some of the terms of a formula published in an article on "The average internal curve, and its application to meteorological phenomena."

Calculating the standard departure of an Ingersoll watch from its normal inaccuracy:

IF:

M = average skewness in minutes,
M = any convenient number near M as for instance 55,
X = difference between Ingersoll time and real time,
N = number of weeks an Ingersoll watch can be expected to run,
D = standard departure of an Ingersoll,
S = full stop

THEN we have

$$D = \sqrt{\dfrac{S(X - 7\,N2)}{M}} = \dfrac{MX = \pm42}{SN}$$

Indicating that the watch is either 42 minutes slow or 42 minutes fast, as may be determined by looking at the sun.

Classes of Farmers and the Ingersoll Habit

If we classify farmers on the Ingersoll watch factor, they fall into two great general classes: first, those who have; and second, those who have not. From a re-examination of the records obtained by Thompson and Dixon in "A farm management survey of three representative areas in Indiana, Illinois and Iowa" it appears that 123 farmers had Ingersolls and 276 had not.

Of the 123 Ingersoll-owning farmers 17 were owners additional, i.e. owned Ingersolls and rented one or more other watches.

Of the remaining Ingersoll owners 3 owned a Chevrolet, 6 used a safety razor, 4 had been classmates of W. J. Bryan, 8 had married young, 19 smoked cigarettes, 8 had protested against fixing the price of wheat, 6 had been operated on for appendicitis, 9 had inherited their farms, 123 were regular attendants at the movies, and one carried a cane on Sundays.

Of the 276 non-Ingersoll-owners all were sorry, 17 had never been in jail, 6 wore celluloid collars, 13 were vaccinated, 7 had tried to make speeches, 9 were caught in the draft, 14 married fat widows with children, 4 had tire blow-outs, 21 were red-headed, 15 had arguments with their neighbors, 10 were fathers of twins, 5 had stomach trouble, and 1 was a Democrat.

The hard luck index is thus seen to be exactly 11% higher with non-owners than with owners.

On the power of the Ingersoll watch to regulate labor income, make children beautiful, sweeten the disposition of wives and neighbors, prevent diseases of man and beasts, we quote from Chester County Bulletin, as follows:

On the 378 farms, 278 were watch owners and of these 277 owned Ingersolls. The 278th man we learned after several subsequent visits promised to purchase a watch of this kind if the Office of Farm Management would prove that it would make him a negative labor income. This was cheerfully done.

Of the 277 owners, 177 rotated their watches from pocket to pocket, while the remaining 100 left their watches at the house and guessed at the time. On an average the latter received $102.05 more labor income per farm than the former, due largely to the wear and tear on the pockets, repairs, etc.

There seems to be a striking relation between Ingersoll watches and adjusted labor income, the adjustment of the watch several times per day serving to adjust the labor income to the needs of the family, the call for cash and the outlay for wearing apparel such as boots, shoes, dresses, gowns, face powders, harness, currycombs, etc.

There is a possibility also of the Ingersoll watch being a preventative of all undesirable ailments and diseases. It was found that of the 177 who carried watches in their pockets had no corns [*sic*] while all but 6 of the remainder had corns on each toe on both feet. It is beyond the province of this bulletin to suggest that carrying a watch, on the same principle of carrying a potato in the pocket, keeps away corns and if corns why not other diseases such as green apple colic, rheumatism, gout, bots, and a thousand other diseases to which human flesh is heir.

A very noticeable feature is that all the children of the watch owners had dimples in their cheeks. This is the first time anything like this has been discovered by farm surveys and the only way to account for it is that the children while yet young listened often to the tick and smiled deeply. It was noticeable also that the disposition of the wives of the owners were sweeter and that life insurance was almost wholely [*sic*] unnecessary.

Genetics of the Ingersoll Watch

Careful records taken by Hawthorne on 76 Ohio farms during the past 4 generations of owners brought out some interesting and hitherto unpublished data on [M]endelism in the ownership of Ingersolls. The passion to possess Ingersolls appears to be a clear case of sex-linked character being inherited only by males without regard to the zygotic composition of either sire or dam. The Ingersoll watch factor was found to be a pure Mendelian factor conforming strictly to the 7-come-11 formula. It is in some mysterious way associated with the positive factors for freckles and fondness for dogs.

The Ingersoll factor showed itself to be dominant in farmers with labor income between $68.31 and $123.67 per year, becoming recessive after the income passed the $7000 mark.

A questionnaire sent to the 76 farmers failed to bring in any important additional information. Seventy-five replied that the Government had no right to make such personal inquiries while one farmer frankly stated that he refused to be the heterozygote in this matter.

Forthcoming Special Bulletins

Drake.   "Hogging down the high cost of living" and "Sheeping off the rent."

Cotton.   "The demobilization of bulls" and "Portable goat pastures."

Wooton.   "The real cause of prickly pears."

Cates.   "Squirrel farming as related to the nut-crop in war time."

Wilcox.   "Farmerettes as laborers in the garden of Allah."

Miller.   "Mule psychology as affected by the 'work or fight' order."

Dixon.   "Breaking point of a farmer's patience as affected by farm surveys."

Bibliography

Archival Materials

Bryan, Enoch Albert. Papers. Manuscripts, Archives, and Special Collections, Holland Library. Washington State University, Pullman.

Spillman, Ramsay. "A Biography of William Jasper Spillman." 1940. Manuscripts, Archives, and Special Collections, Holland Library. Washington State University, Pullman.

Spillman, William J. Papers. Manuscripts, Archives, and Special Collections, Holland Library. Washington State University, Pullman.

———. Papers. National Library of Agriculture. U.S. Department of Agriculture, Beltsville, MD.

———. Papers. Western Historical Manuscript Collection. University of Missouri–Columbia Archives.

U.S. Department of Agriculture Archives. National Archives and Records Collection, College Park, MD.

Newspapers

Bangor (ME) Daily Commercial
Coeur d'Alene (ID) Independent
New York Globe and Commercial Advertiser
New York Times
Pullman (WA) Herald
Spokane (WA) Spokesman-Review

Government Documents

Bryan, Enoch A. *Second Annual Report: Washington Agricultural College and School of Science.* Olympia: O. C. White, State Printer, 1893.

————. *Fifth Annual Report: Washington Agricultural Experiment Station, Washington State Department of Agriculture.* Olympia: O. C. White, State Printer, 1896.

————. *Sixth Annual Report: Washington Agricultural Experiment Station, Washington State Department of Agriculture.* Olympia: O. C. White, State Printer, 1897.

Congressional Record. 1913–1916. Washington, DC.

Elliott, E. E. *Twelfth Annual Report of the Director of the Washington Agricultural Experiment Station: Washington State Department of Agriculture.* Olympia: O. C. White, State Printer, 1902.

Secretary of Agriculture. *Annual Reports of the Department of Agriculture: Fiscal Year Ended June 30, 1902.* Washington, DC: Government Printing Office, 1902.

Other Sources

Adams, Mark, ed. *The Wellborn Science: Eugenics in Germany, France, Brazil, and Russia.* New York: Oxford University Press, 1990.

Albertson, Dean. *Roosevelt's Farmer: Claude R. Wickard in the New Deal.* New York: Columbia University Press, 1961.

"Alcohol as a Motor Fuel." *Engineering Magazine* (April 1904). From collection of Henry Ford Museum.

American Breeders' Association. *Proceedings.* Vol. 1. Washington, DC: American Breeders' Association, 1905.

————. *Proceedings.* Vol. 2. Washington, DC: American Breeders' Association, 1906.

————. *Proceedings.* Vol. 6. Washington, DC: American Breeders' Association, 1909.

American Country Life Association. *Farm Income and Farm Life: A Symposium on the Relation of the Social and Economic Factors in Rural Progress.* Chicago: University of Chicago Press for the American Country Life Association, New York, 1927.

"American Farmers! Farm Efficiency Is Essential." *Tractor Farming* (April 1919): 2.

Anderson, Oscar. *Health of a Nation.* Chicago: University of Chicago Press, 1958.

Ankli, Robert E. "Horses vs. Tractors on the Corn Belt." *Agricultural History* 54 (Spring 1980): 134–48.

Apple, Michael W., and Linda K. Christian-Smith, eds. *The Politics of the Textbook*. New York: Routledge, 1991.

Bailey, Liberty Hyde. *Cyclopedia of American Agriculture: A Popular Survey of Agricultural Conditions, Practices, and Ideals in the United States and Canada*. Vol. 1. New York: Macmillan, 1909.

———. *Plant-Breeding: Being Six Lectures upon the Amelioration of Domestic Plants*. New York: Macmillan, 1895.

Baker, Gladys L., Wayne D. Rasmussen, Vivian Wiser, and Jane M. Porter. *Century of Service: The First 100 Years of the United States Department of Agriculture*. Washington, DC: U.S. Department of Agriculture, Economic Research Service, 1963.

Benedict, Murray R. *Farm Policies of the United States, 1790–1950: A Study of Their Origins and Development*. New York: Twentieth Century Fund, 1953.

Berger, Samuel R. *Dollar Harvest: The Story of the Farm Bureau*. Lexington, MA: Heath Lexington Books, 1971.

Block, William Joseph. *The Separation of the Farm Bureau and the Extension Service*. Urbana: University of Illinois Press, 1960.

Bonner, Thomas Neville. *Iconoclast: Abraham Flexner and a Life in Learning*. Baltimore: Johns Hopkins University Press, 2002.

Bowen, Eugene R. *Tractor Farming*. Peoria, IL: Avery, 1914.

Brachvogel, John Kudlich. *Industrial Alcohol, Its Manufacture and Uses: A Practical Treatise Based on Dr. Max Maercker's "Introduction to Distillation" as Revised by Dr. Delbruck and Dr. Lange, Comprising Raw Materials, Malting, Mashing and Yeast Preparation, Fermentation, Distillation, Rectification, and Purification of Alcohol, a Statistical Review, and the United States Law*. New York: Munn, 1907.

Bressman, E. N. "Spillman's Work on Plant Breeding." *Science* (September 23, 1932): 273–74.

Campbell, Christiana McFadyen. *The Farm Bureau and the New Deal: A Study of the Making of National Farm Policy, 1933–40*. Urbana: University of Illinois Press, 1962.

Carstensen, Vernon. "A Century of the Land-Grant Colleges." *Journal of Higher Education* (January 1962): 30–37.

Carver, T. N., H. C. Taylor, and G. F. Warren. "William Jasper Spillman, 1863–1931: First President of the American Farm Management Association." *Journal of Farm Economics* 14 (January 1932): 1–22.

Chernow, Ron. *Titan: The Life of John D. Rockefeller, Sr.* New York: Random House, 1998.

"Consumption of Fertilizers in the Southern States." *American Fertilizer* (Philadelphia) (January 31, 1920): 134.

Coppin, Clayton, and Jack High. *The Politics of Purity: Harvey Washington Wiley and the Origins of Federal Food Policy.* Ann Arbor: University of Michigan Press, 1999.

Crawford, Harriet Ann. *The Washington State Grange, 1889–1924: A Romance of Democracy.* Portland: Binfords and Mort, 1940.

Cremin, Lawrence A. *The Transformation of the School: Progressivism in American Education, 1876–1957.* New York: Alfred A. Knopf, 1961.

Cronon, William. *Nature's Metropolis: Chicago and the Great West.* New York: W. W. Norton, 1991.

Cullen, George A. *The Cradle of the Farm Bureau Idea: Marketing Possibilities of the Bureau.* Binghamton, NY: Broome County Farm Bureau, 1920.

———. "The Cradle of the Farm Bureau Idea: Marketing Possibilities of the Bureau." Speech given at annual meeting of Broome County Farm Bureau, Binghamton, NY, 1920.

Culver, John C., and John Hyde. *American Dreamer: The Life and Times of Henry A. Wallace.* New York: W. W. Norton, 2000.

Danbom, David B. "The Agricultural Experiment Station and Professionalization: Scientists' Goals for Agriculture." *Agricultural History* 60 (Spring 1986): 246–55.

———. *Born in the Country: A History of Rural America.* Baltimore: Johns Hopkins University Press, 1995.

———. *The Resisted Revolution: Urban America and the Industrialization of Agriculture, 1900–1930.* Ames: Iowa State University Press, 1979.

Davenport, Charles R. "Determination of Dominance in Mendelian Inheritance." *Proceedings of the American Philosophical Society* 47 (1908): 59–63.

Davis, J. S., Helen M. Gibbs, and Elizabeth Brand Taylor. *Wheat in the World Economy: A Guide to Wheat Studies of the Food Research Institute.* Stanford: Food Research Institute, 1945.

Dean, William Harper. "What's the Matter with the Department?" *Country Gentleman* 82 (March 3, 1917): 415–16.

———. "What's the Matter with the Department?" *Country Gentleman* 82 (March 17, 1917): 8–9.

———. "What's the Matter with the Department? The Confusion of Tongues." *Country Gentleman* (March 24, 1917): 13–14.

Derry, T. K., and Trevor I. Williams. *A Short History of Technology: From the Earliest Times to A.D. 1900.* 1963. Reprint, New York: Dover Publications, 1993.

Dictionary of American History. Vol. 1. Rev. ed. New York: Charles Scribner's Sons, 1976.

Dixon, D. F. "The Growth of Competition among the Standard Oil Companies in the United States, 1911–1961." *Business History* 9 (1967): 1–29.

Duesterhaus, Richard. "Sustainability's Promise." *Journal of Soil and Water Conservation* 45 (January–February 1990): 4.

Dunn, L. C. "The *American Naturalist* in American Biology." *American Naturalist* 100 (September–October 1966): 481–92.

Dupré, Ruth. "'If It's Yellow, It Must Be Butter': Margarine Regulation in North America since 1886." *Journal of Economic History* 59 (June 1999): 353–71.

Dupree, A. Hunter. *Science in the Federal Government: A History of Policies and Activities to 1940.* New York: Harper and Row, 1957.

"The Editor's Viewpoints." *Farm Journal* (October 1918).

Elliott, E. E. "The Farmer's Changing World." In *Farmers in a Changing World: U.S. Department of Agriculture Yearbook.* Washington, DC: Government Printing Office, 1940.

Ellis, L. W., and Edward A. Rumely. *Power and the Plow.* Garden City, NY: Doubleday, Page, 1911.

Etzioni-Halevy, Eva. *Bureaucracy and Democracy: A Political Dilemma.* Boston: Routledge and Kegan Paul, 1983.

Fahey, John. *The Inland Empire: Unfolding Years, 1879–1929.* Seattle: University of Washington Press, 1986.

Fairchild, David. *The World Was My Garden: Travels of a Plant Explorer.* New York: Charles Scribner's Sons, 1938.

Ferleger, Louis. "Arming American Agriculture for the Twentieth Century: How the USDA's Top Managers Promoted Agricultural Development." *Agricultural History* 74 (Spring 2000): 211–26.

Fite, Gilbert C. *American Farmers: The New Minority.* Bloomington: Indiana University Press, 1981.

———. *George N. Peek and the Fight for Farm Parity.* Norman: University of Oklahoma Press, 1954.

Fitzgerald, Deborah. *Every Farm a Factory: The Industrial Ideal in American Agriculture.* New Haven: Yale University Press, 2003.

Fosdick, Raymond B. *Adventure in Giving: The Story of the General Education Board.* New York: Harper and Row, 1962.

Garraty, John A., and Mark C. Carnes, eds. *American National Biography.* New York: Oxford University Press, 1999.

The General Education Board: An Account of Its Activities, 1902–1914. New York: General Education Board, Rockefeller Foundation, 1915.

General Education Board: Review and Final Report, 1902–1964. New York: General Education Board, Rockefeller Foundation, 1964.

George, Alexander L., and Juliette L. George. *Woodrow Wilson and Colonel House: A Personality Study.* New York: John Day, 1956.

"Getting Down to Brass Tacks." *Farm Journal* (January 1920): 8.

Giebelhaus, August W. "Farming for Fuel: The Alcohol Motor Fuel Movement of the 1930s." *Agricultural History* 54 (Spring 1980): 173–84.

Goodwyn, Lawrence. *Democratic Promise: The Populist Moment in America.* New York: Oxford University Press, 1976.

Gould, Lewis L. *Progressives and Prohibitionists: Texas Democrats in the Wilson Era.* Austin: University of Texas Press, 1973.

Hamilton, David E. *From New Day to New Deal: American Farm Policy from Hoover to Roosevelt, 1928–1933.* Chapel Hill: University of North Carolina Press, 1991.

Hamilton, J. *History of Farmers' Institutes in the United States.* Bulletin 174. Washington, DC: U.S. Department of Agriculture, Office of Experiment Stations, 1906.

Harden, Victoria. *Rocky Mountain Spotted Fever: History of a Twentieth-Century Disease.* Baltimore: Johns Hopkins University Press, 1990.

Harding, T. Swann. *Two Blades of Grass: A History of Scientific Development in the United States Department of Agriculture.* Norman: University of Oklahoma Press, 1947.

Hawley, Ellis W. "Herbert Hoover, the Commerce Secretariat, and the Vision of the 'Associative State.'" *Journal of American History* 61 (June 1974): 116–40.

Hays, Samuel P. *The Response to Industrialism: 1885–1914.* Chicago: University of Chicago Press, 1957.

Hays, Willet M. "Breeding Problems." In *Proceedings: American Breeders' Association,* 1:197. Washington, DC, 1905.

Hayter, Earl W. *The Troubled Farmer, 1900–1950: Rural Adjustment to Industrialism.* Dekalb: Northern Illinois University Press, 1968.

Hendrick, Burton J. *The Life and Letters of Walter H. Page.* 3 vols. Garden City, NY: Doubleday, Page, 1922–1925.

Hendricks, Walter A., Morley A. Jull, and Harry W. Titus. "A Possible Physiological Interpretation of the Law of the Diminishing Increment." *Science* (April 17, 1931): 427–29.

Henig, Robin Marantz. *The Monk in the Garden: The Lost and Found Genius of Gregor Mendel, the Father of Genetics.* Boston: Houghton Mifflin, 2000.

Herrick, Rufus Frost. *Denatured or Industrial Alcohol: A Treatise on the History, Manufacture, Composition, Uses, and Possibilities of Industrial Alcohol in the Various Countries Permitting Its Use, and the Laws and Regulations Governing the Same, Including the United States.* New York: J. Wiley and Sons, 1907.

Hillison, John. "Agricultural Education and Cooperative Extension: The Early Agreements." *Journal of Agricultural Education* 37 (Spring 1996).

Hodge, Clifton F., and Jean Dawson. *Civic Biology: A Textbook of Problems, Local and National, That Can Be Solved Only by Civic Cooperation.* Boston: Ginn, 1918.

Hoing, Willard Lee. "James Wilson as Secretary of Agriculture, 1897–1913." PhD diss., University of Wisconsin, 1964.

Holmes, S. J. *Human Genetics and Its Social Import.* New York: McGraw-Hill, 1936.

Hoover, Herbert. "A New World Food Situation." *Banker-Farmer* 6 (December 1918): 4–5.

Houston, David Franklin. *Eight Years with Wilson's Cabinet, 1913–1920, with a Personal Estimate of the President.* Garden City, NY: Doubleday, Page, 1926.

Howard, George E. "The State University in America." *Atlantic Monthly* 67 (March 1891): 332–53.

Howard, Robert P. *James R. Howard and the Farm Bureau.* Ames: Iowa State University Press, 1983.

Huber, Richard M. *The American Idea of Success.* New York: McGraw-Hill, 1971.

Hurst, Charles Chamberlain. *Experiments in Genetics.* London: Cambridge University Press, 1925.

Hurt, R. Douglas. *Problems of Plenty: The American Farmer in the Twentieth Century.* Chicago: Ivan R. Dee, 2002.

Hutson, Cecil Kirk. "Texas Fever in Kansas, 1866–1930." *Agricultural History* 68 (Spring 1994): 74–104.

Jaworski, Janusz. "Decision Aspects of the Spillman Production Function." *Canadian Journal of Agricultural Economics* 25 (1977): 48–53.

Jones, Stephen S., and Molly M. Cadle. "Spillman, Gaines, and Vogel— Building a Foundation." *Wheat Life* (February 1996): 25–34.

"Junior Farmers." *Tractor Farming* (April 1919): 9.

Kelly, Fred C. "A Wonderful Question Answerer." *American Magazine* (n.d.): 47–48.

Kimmelman, Barbara A. "The American Breeders' Association: Genetics and Eugenics in an Agricultural Context, 1903–13." *Social Studies of Science* 13 (1983): 163–204.

———. "A Progressive Era Discipline: Genetics at American Agricultural Colleges and Experiment Stations, 1900–1920." PhD diss., University of Pennsylvania, 1987.

King, F. H. *Farmers of Forty Centuries; or, Permanent Agriculture in China, Korea, and Japan.* Madison: Mrs. F. H. King, 1911.

Kirkendall, Richard S. "The Agricultural Colleges: Between Tradition and Modernization." *Agricultural History* 60 (Spring 1986): 3–21.

Knapp, S. A. *Demonstration Work in Southern Farms.* Farmers' Bulletin 422. Washington, DC: U.S. Department of Agriculture, 1910.

Kulikoff, Allan. *The Agrarian Origins of American Capitalism.* Charlottesville: University Press of Virginia, 1992.

Lacey, Robert. *Ford: The Men and the Machine.* Boston: Little, Brown, 1986.

Lamer, Mirko. *The World Fertilizer Economy.* Stanford: Stanford University Press, 1957.

National Association of State Universities and Land-Grant Colleges. *The Land Grant Tradition.* Washington, DC: National Association of State Universities and Land-Grant Colleges, 1995.

Lang, Charles Louis. "A Historical Review of the Forces That Contributed to the Formation of the Cooperative Extension Service." PhD diss., Michigan State University, 1975.

Lasch, Christopher. *The Revolt of the Elites and the Betrayal of Democracy.* New York: W. W. Norton, 1995.

Lewis, David Levering. *W. E. B. Du Bois: The Fight for Equality and the American Century, 1919–1963.* New York: Henry Holt, 2000.

Lewis, Eugene. *Public Entrepreneurship: Toward a Theory of Bureaucratic Political Power.* Bloomington: Indiana University Press, 1980.

Lord, Russell. *The Wallaces of Iowa.* Boston: Houghton Mifflin, 1947.

Ludmerer, Kenneth M. *Genetics and American Society: A Historical Approach.* Baltimore: Johns Hopkins University Press, 1972.

MacDowell, Charles H. "Problems and Processes in Mixed Fertilizers." *American Fertilizer* (Philadelphia) (February 14, 1920): 67–72.

Macmahon, Arthur W. "Selection and Tenure of Bureau Chiefs in the National Administration of the United States." *American Political Science Review* 20 (August 1926): 548–82.

MacPhail, Sir Andrew. *Three Persons.* London: John Murray, 1929.

Marcus, Alan I. *Agricultural Science and the Quest for Legitimacy: Farmers, Agricultural Colleges, and Experiment Stations, 1870–1890.* Ames: Iowa State University Press, 1985.

———. "The Ivory Silo: Farmer-Agricultural College Tensions in the 1870s and 1880s." *Agricultural History* 60 (Spring 1986): 22–36.

Mayberry, B. D. "The Tuskegee Movable School: A Unique Contribution to National and International Agriculture and Rural Development." *Agricultural History* 65 (Summer 1991): 85–104.

McClelland, Peter D. *Sowing Modernity: America's First Agricultural Revolution.* Ithaca: Cornell University Press, 1997.

McConnell, Grant. *The Decline of Agrarian Democracy.* Berkeley and Los Angeles: University of California Press, 1953.

McCoy, Joseph G. *Historic Sketches of the Cattle Trade of the West and Southwest.* 1874. Reprint, Washington, DC: Rare Book Shop, 1932.

McDowell, George R. *Land-Grant Universities and Extension into the 21st Century.* Ames: Iowa State University Press, 2001.

McGeary, M. Nelson. *Gifford Pinchot: Forester-Politician.* Princeton: Princeton University Press, 1960.

Meade, Melinda S., John W. Florin, and Wilbert M. Gesler. *Medical Geography.* New York: Guilford Press, 1988.

Meindl, Christopher F., Derek H. Alderman, and Peter Waylen. "On the Importance of Environmental Claims-Making: The Role of James O. Wright in Promoting the Drainage of Florida's Everglades in the Early Twentieth Century." *Annals of the Association of American Geographers* 92 (2002): 682–701.

Miller, Char. *Gifford Pinchot and the Making of Modern Environmentalism.* Washington, DC: Island Press, 2001.

Moore, Sherrel. "From Progressive to Radical: William and Lura Bouck in the Washington State Grange." Master's thesis, Western Washington University, 1992.

Morris, Andrew. "The General Education Board and the U.S.D.A." *Research Reports from the Rockefeller Archive Center* (Sleepy Hollow, NY) (Spring 1999): 16–19.

National Grange. *Journal of Proceedings.* Washington, DC: National Grange, 1905, 1907–1908.

Neth, Mary. *Preserving the Family Farm: Women, Community, and the Foundations of Agribusiness in the Midwest, 1900–1940.* Baltimore: Johns Hopkins University Press, 1995.

Nevins, Allan. *The State Universities and Democracy.* Urbana: University of Illinois Press, 1962.

Nitrogen. New York: Synthetic Nitrogen Products Association, 1924.

Nordin, D. Sven. *Rich Harvest: A History of the Grange, 1867–1900.* Jackson: University Press of Mississippi, 1974.

Norris, Frank. *The Octopus: A Story of California.* Garden City, NY: Doubleday, Page, 1901.

Olby, Robert C. *Origins of Mendelism.* New York: Schocken Books, 1966.

Oleson, Alexandra, and John Voss, eds. *The Organization of Knowledge in Modern America, 1860–1920.* Baltimore: Johns Hopkins University Press, 1979.

Otsuka, Keijiro, Hiroyuki Chuma, and Yujiro Hayami. "Land and Labor Contracts in Agrarian Economies: Theories and Facts." *Journal of Economic Literature* 30 (1992): 1965–2018.

"The Outlook for Agriculture." *Banker-Farmer* 6 (December 1918): 2.

Pagé, Victor W. *The Modern Gas Tractor: Construction, Utility, Operation, and Repair.* New York: Norman W. Henley, 1922.

Painter, Nell Irvin. *Standing at Armageddon: The United States, 1877–1987.* New York: W. W. Norton, 1987.

Pawson, H. Cecil. *Robert Bakewell: Pioneer Livestock Breeder.* London: Crosby, Lockwood, and Son, 1957.

Perkins, John H. *Geopolitics and the Green Revolution: Wheat, Genes, and the Cold War.* New York: Oxford University Press, 1997.

Perkins, Van L. *Crisis in Agriculture: The Agricultural Adjustment Administration and the New Deal, 1933.* Berkeley and Los Angeles: University of California Press, 1969.

Pinchot, Gifford. *The Power Monopoly: Its Make-Up and Its Menace.* Milford, PA: by the author, 1928.

The Potash Industry. Chicago: German Kali Works, 1912.

Ramaley, Francis. "Mendelian Proportions and the Increase of Recessives." *American Naturalist* 46 (June 1912): 344–51.

Rasmussen, Wayne D., ed. *Agriculture in the United States: A Documentary History.* 4 vols. New York: Random House, 1975.

———. *Readings in the History of American Agriculture.* Urbana: University of Illinois Press, 1960.

Rasmussen, Wayne D., and Gladys L. Baker. *The Department of Agriculture.* New York: Praeger, 1972.

Reid, Joseph D., Jr. Review of *Agriculture in the United States: A Documentary History,* ed. Wayne D. Rasmussen. *Journal of Economic History* 37 (1977): 553–54.

Rice, Thurman B. *Racial Hygiene: A Practical Discussion of Eugenics and Race Culture.* New York: Macmillan, 1929.

Rife, David C. *Hybrids.* Washington, DC: Public Affairs Press, 1965.

Rodgers, Daniel T. *Atlantic Crossings: Social Politics in a Progressive Age.* Cambridge: Harvard University Press, 1998.

Rommel, George M. *Essentials of Animal Breeding.* Farmers' Bulletin 1167. Washington, DC: U.S. Department of Agriculture, 1920.

———. *Farm Products in Industry.* New York: Rae D. Henkle, 1928.

Rosenberg, Charles E. *No Other Gods: On Science and American Social Thought.* Baltimore: Johns Hopkins University Press, 1997.

Ross, A. B. "Commercial Nitrogen—the Great Gold Brick." *Farm Journal* (November 1919): 8–12.

———. "Old Fertilizer Theories Scrapped." *Farm Journal* (October 1919): 10–12.

Ross, Earle Dudley. *Democracy's College: The Land-Grant Movement in the Formative Stage.* Ames: Iowa State College Press, 1942.

Rossiter, Margaret. "The Organization of the Agricultural Sciences." In *The Organization of Knowledge in Modern America, 1860–1920,* ed. Alexandra Oleson and John Voss, 211–48. Baltimore: Johns Hopkins University Press, 1979.

Rowley, William D. "M. L. Wilson: Believer in the Domestic Allotment." *Agricultural History* 43 (April 1969): 277–87.

———. *M. L. Wilson and the Campaign for the Domestic Allotment.* Lincoln: University of Nebraska Press, 1970.

Rumeley, Edward A. "The Passing of the Man with the Hoe." *World's Work* 20 (August 1910): 13246–58.

Saloutos, Theodore. *The American Farmer and the New Deal.* Ames: Iowa State University Press, 1982.

Saloutos, Theodore, and John D. Hicks. *Agricultural Discontent in the Middle West: 1900–1939.* Madison: University of Wisconsin Press, 1951.

Sandoz, Mari. *Old Jules.* 1935. Reprint, Lincoln: University of Nebraska Press, 1985.

Schmidt, Louis Bernard, and Earle Dudley Ross. *Readings in the Economic History of American Agriculture.* New York: Macmillan, 1925.

Schuler, Loring, ed. *The "Country Gentleman": An Analysis of Content.* Philadelphia: Curtis Publishing, 1925.

Schwantes, Carlos A. *Coxey's Army: An American Odyssey.* Lincoln: University of Nebraska Press, 1985.

Scobie, James R. *Revolution on the Pampas: A Social History of Argentine Wheat, 1860–1910.* Austin: University of Texas Press, 1964.

Scott, Roy Vernon. *The Reluctant Farmer: The Rise of Agricultural Extension to 1914.* Urbana: University of Illinois Press, 1970.

Sealander, Judith. *Private Wealth and Public Life: Foundation Philanthropy and the Reshaping of American Social Policy from the Progressive Era to the New Deal.* Baltimore: Johns Hopkins University Press, 1997.

Shull, George Harrison. "The 'Presence and Absence' Hypothesis." *American Naturalist* 43 (July 1909): 410–19.

Sinclair, Upton. *The Goose-Step: A Study of American Education.* Pasadena: by the author, 1923.

Smith, Arthur D. Howden. *Mr. House of Texas.* New York: Funk and Wagnalls, 1940.

Smith, Joe. *Bunch Grass Pioneer.* Fairfield, WA: Ye Galleon Press, 1986.

Southern Fertilizer Association. *Yearbook.* Atlanta: Southern Fertilizer Association, 1924, 1925.

Squires, H. G. "Alcohol Motors and Pumps in Cuba: August 20, 1904." *U.S. Monthly Consular Reports* (September 1904).

Stevenson, James Henry. *Traction Farming and Traction Engineering: Gasoline, Alcohol, Kerosene; A Practical Hand-Book for the Owners and Operators of Gas and Oil Engines on the Farm.* Chicago: F. J. Drake, 1915.

Stockbridge, Frank Parker. "A University That Runs a State." *World's Work* (April 1913): 699–708.

Strom, Claire. "Texas Fever and the Dispossession of the Southern Yeoman Farmer." *Journal of Southern History* 66 (Spring 2000): 49–74.

Tarbell, Ida M. *The History of the Standard Oil Company.* 2 vols. New York: McClure, Phillips, 1904.

Taylor, Frederick W. *The Principles of Scientific Management.* 1911. Reprint, Mineola, NY: Dover Publications, 1998.

Taylor, Henry C., and Anne Dewees Taylor. *The Story of Agricultural Economics in the United States, 1840–1932.* Ames: Iowa State College Press, 1952.

Ten Eyck, A. M. *Wheat.* Lincoln: Campbell Soil Culture Publishing, 1914.

Thomson, E. H. "The Origin and Development of the Office of Farm Management in the United States Department of Agriculture." *Journal of Farm Economics* 14 (January 1932): 11–16.

Townley, A. C. "Is the National Nonpartisan League the Answer?" *Farm Journal* (October 1918): 8.

———. "The Problem of the West." *Atlantic Monthly* (September 1896).

True, Alfred Charles. "Five Years of the Smith-Lever Extension Act: County Agent Movement." *American Farming* (Chicago) (1919).

———. *A History of Agricultural Experimentation and Research in the United States, 1607–1925.* U.S. Department of Agriculture, Miscellaneous Publication 251. Washington, DC: Government Printing Office, 1937.

———. *A History of Agricultural Extension Work in the United States, 1785–1923.* U.S. Department of Agriculture, Miscellaneous Publication 15. Washington, DC: Government Printing Office, 1928.

Turner, Frederick Jackson. "Dominant Forces in Western Life." *Atlantic Monthly* 78 (April 1897): 433–43.

———. *The Frontier in American History.* New York: Henry Holt, 1920.

Unger, Harlow Giles. *Noah Webster: The Life and Times of an American Patriot.* New York: John Wiley and Sons, 1998.

"University and Educational News." *Science,* n.s., 46 (December 28, 1917): 638.

Viereck, George Sylvester. *The Strangest Friendship in History: Woodrow Wilson and Colonel House.* 1932. Reprint, Westport, CT: Greenwood Press, 1976.

Wallace, Henry A. *Democracy Reborn.* Ed. Rusisell Lord. New York: Reynal and Hitchcock, 1944.

———. *New Frontiers.* New York: Reynal and Hitchcock, 1934.

Warwick, Everett J. "New Breeds and Types." In *Yearbook of Agriculture,* 276–80. Washington, DC: Government Printing Office, 1962.

White, Richard. *It's Your Misfortune and None of My Own.* Norman: University of Oklahoma Press, 1991.

Wiebe, Robert H. *The Search for Order, 1877–1920.* New York: Hill and Wang, 1967.

Wik, Reynold M. *Henry Ford and Grass-Roots America.* Ann Arbor: University of Michigan Press, 1972.

Wilcox, Earley Vernon. *Tama Jim.* Boston: Stratford, 1930.

Wiley, Harvey W. *The History of a Crime against the Food Law.* Washington, DC: by the author, 1929.

Williams, Robert C. *Fordson, Farmall, and Poppin' Johnny: A History of the Farm Tractor and Its Impact on America.* Urbana: University of Illinois Press, 1987.

Willis, H. Parker. "Secretary Wilson's Record." *Collier's* (March 23, March 30, and April 6, 1912).

Winters, Donald L. *Henry Cantwell Wallace, as Secretary of Agriculture, 1921–1924.* Urbana: University of Illinois Press, 1970.

Woods, Frederick Adams. "American Men of Science and the Question of Heredity." *Science* (August 13, 1909): 205–9.

———. "The Birthplaces of Leading Americans and the Question of Heredity." *Science* (July 2, 1909): 17–21.

———. "City Boys versus Country Boys." *Science* (April 9, 1909): 577–79.

———. "Sovereigns and the Supposed Influence of Opportunity." *Science* (June 19, 1914): 902–5.

Woodward, C. Vann. *Origins of the New South, 1877–1913.* Baton Rouge: Louisiana State University Press, 1951.

Worster, Donald. *Dust Bowl: The Southern Plains in the 1930s.* New York: Oxford University Press, 1979.

Wright, David E. "Alcohol Wrecks a Marriage: The Farm Chemurgic Movement and the USDA in the Alcohol Fuels Campaign in the Spring of 1933." *Agricultural History* 67 (Spring 1993): 36–66.

Yamazaki, W. T., and C. T. Greenwood, eds. *Soft Wheat: Production, Breeding, Milling, and Uses.* St. Paul: American Association of Cereal Chemists, 1981.

Yergin, Daniel. *The Prize: The Epic Quest for Oil, Money, and Power.* New York: Simon and Schuster, 1992.

Yerkes, Arnold P., and H. H. Mowry. *Farm Experience with the Tractor.* Bulletin 174. Washington, DC: U.S. Department of Agriculture, 1915.

Index

Agricultural Adjustment Administration, 152, 153, 158
Agricultural Extension Service, 34, 65
Agricultural ladder, 45
Agricultural Marketing Act of 1929, 146
American Association of Agricultural Colleges and Experiment Stations, 20, 24, 66, 67
American Association for Agricultural Legislation, 91
American Bankers Association, 117
American Breeders' Association, 24, 25
American Economic Association, 89, 90, 91
American Farm Economic Association, 154
American Federation of Labor, 125
American Naturalist, 28–29, 86
American Society of Equity, 24, 123, 125
Arnold, J. H., 96–97
Atavism, 12, 26

Bailey, Liberty Hyde, 75
Bangor, Maine, 136
Bateson, William, 11, 29
Better Farming Association, 125
Black, John D., 150–51
Bolshevism: threat of, 118, 125–27
Broome County, New York, 54, 124
Bryan, Enoch A., 15, 21, 155–56
Bureau of Agricultural Economics: and Extension Service, 124; Spillman at, 136
Bureau of Animal Industry, 95, 114
Bureau of Chemistry, 95

Bureau of Plant Industry, 23; records destroyed, 158

Carver, George Washington, 43
Cates, J. S., 53
Cattelo, 26
Censorship: of USDA reports, 88, 89, 95
Center for Sustaining Agricultural and Natural Resources, 159
Chambers of Commerce, 124, 125, 136, 155
Chautauqua movement, 38, 40, 71
Chilean nitrates, 102, 105–6
Cold Springs Harbor, New York, 27, 28
Cooperative Extension Service, 67, 71, 159
Cooperative marketing, 123, 124, 127; in Ireland, 128; failure of, 143; in California, 145
Cornell University, 108, 124
Correns, Karl, 11, 20, 29
Cost of production, 97, 130, 134, 139–40
Cotton boll weevil, 56, 107
Cotton farmers: and fertilizer usage, 105, 107, 110, 113
Country Gentleman magazine, 76, 79, 80
Country Life Commission, 32, 127
County agent model, 55
Crampton, Louis, 94
Cropland, dormant, 144, 147

Dairymen's League, 94
Darwin, Charles, 10, 19, 22, 32–33
Davenport, Charles, 27, 29